# THERMAL SHUTTERS
# AND SHADES

# THERMAL SHUTTERS AND SHADES

### Over 100 Schemes for
### Reducing Heat-Loss Through Windows

## WILLIAM A. SHURCLIFF

**BRICK HOUSE PUBLISHING CO., INC.**
**Andover, Massachusetts**

Published by Brick House Publishing Co., Inc.
  3 Main Street
  Andover, Massachusetts 01810

Production credits
  Copy editor: Sandra Hughes Minkkinen
  Book design: Herb Caswell
  Cover design: Susan Slovinsky
  Typesetting: Neil Kelley
  Production supervision: Dixie Clark

Printed in the United States of America

Library of Congress Cataloging in Publication Data

Shurcliff, William A.
  Thermal shutters and shades.

  Indludes index.
  1. Blinds.  2. Window shades.  3. Buildings
—Energy conservation.  I. Title.
TH2276.S58     684      80-11574
ISBN 0-931790-08-5 cloth
ISBN 0-931790-14-X paper

# CONTENTS

## 17 INDOOR OPAQUE ONE-SHEET ROLL-UP SHADES AND ROLL-UP SHUTTERS    143

## 18 INDOOR OPAQUE MULTI-SHEET ROLL-UP SHADES    163

## 19 INDOOR DEVICES OF OTHER TYPES    172

## 20 DEVICES FOR SPECIAL APPLICATIONS    184

# ACKNOWLEDGMENTS

In preparing this book I have been helped—directly or indirectly—by so many people that to list their names would take up too much space. My greatest indebtedness is to Stewart T. Coffin, Steven C. Baer, and John I. Yellott. I am obliged to John C. Gray and Charles H. Shurcliff for criticizing the early drafts and making large numbers of constructive criticisms.

# THERMAL SHUTTERS AND SHADES

# CHAPTER 1

# INTRODUCTION

- Heat-Loss Through Windows Must Be Reduced!
- The Solution: Thermal Shutters and Shades
- Additional Benefits
- The Writer's Goal
- Definitions of Shutters, Shades, Etc.
- Is There Some One Best Choice of Shutter or Shade?
- Use of a Mixture of Strategies
- Functions of a Window
- Main types of Windows
- Conventions Used in Descriptions and Drawings
- Warnings

## HEAT-LOSS THROUGH WINDOWS MUST BE REDUCED!

In typical houses in the northern half of the United States about 15 to 35% of the total heat-loss in winter is via the windows.* In well insulated houses that have double-glazed windows the heat-loss is only about half as great, but the percentage attributable to the windows is presumably about the same: 15 to 35%. In passively solar-heated houses the percentage of heat lost via the windows is greater—about 20 to 40%—because such houses have especially large window areas.

Clearly, the grand total heat-loss-through-windows of U. S. houses is excessive. The loss corresponds to about 300 million barrels of oil per year, or about 3% of our total annual use of purchased energy of all kinds.

As F. M. Schmidt has said (S—40): "The wolf isn't at the door. He is at the window."

## THE SOLUTION: THERMAL SHUTTERS AND SHADES

The crux of the problem is heat-loss at *night* because in winter the nights are longer and colder than the days.

The solution? Develop cheap, thermally effective shutters and shades for use at night. Recent calculations made by Watson et al. (W-110) indicate that the application of high quality (R-15) shades on the windows of an average house with good insulation and a modest window area will reduce the winter heat-need by about 10%. If the house has an especially large area of south windows (364 ft$^2$, i.e., 26% of floor area), use of such shades will reduce the heat-need by about 20%.

If durable, easily operated, attractive shutters and shades are developed, and if they are inexpen-sive, they may be installed within a few years in millions of houses. If they were to be installed in all houses, the overall saving might amount to 50 to 70% of the present overall heat-loss through windows and might reduce our country's total annual usage of purchased energy by 1½ to 2%.

## ADDITIONAL BENEFITS

Besides reducing heat-loss-by-conduction-and-radiation, thermal shutters and shades may reduce heat-loss indirectly by:

Reducing in-leak of cold outdoor air and out-leak of warm indoor air, if the shutters and shades are well sealed.

Reducing cold drafts near the windows.

Raising the general radiation temperature of the room. I am told that if the windows are single glazed and their area is 12% of the floor area of the room, and if the outdoor temperature is 0°F, employment of highly thermally effective shutters and shades increases the radiation temperature of the room sufficiently so that the thermostat may be set about three degrees lower.

Permitting use of higher humidity in the rooms without incurring risk of moisture accumulation on windows. This permits further slight reduction in thermostat setting with no sacrifice of comfort.

Some other benefits are:

Helping keep the house above freezing (above 32°F) in the event of complete cut-off of conventional heating. If a house becomes colder than this, water-filled pipes and tanks may freeze and burst. If thermal shutters and shades can avert this tragedy, they will have paid for themselves many times over.

Some kinds of shutters and shades are highly effective at excluding solar radiation in summer and thus helping keep the house cool.

Some kinds of outdoor shutters can be used during the daytime in winter to reflect additional solar radiation to the windows and thus

---

* The figures presented here are estimates based on a variety of figures presented in the literature of solar heating. As an example of the estimates available, the following paragraph from the Department of Energy's *Energy Insider* of July 23, 1979, may be cited: "The Window Shade News bureau in New York City, a trade organization, reports that one fourth of the energy used in the United States for heating and cooling is squandered through windows; that from 40 cents to $1.40 is added to fuel and utility bills annually for each square foot of single pane glass in a typical home."

increase the extent of solar heating. The benefit may be large: at a typical location in the northern U. S. the annual benefit may amount to about 35¢/ft$^2$ relative to the use of $1-per-gallon oil.

Some kinds of outdoor shutters help exclude burglars.

## THE WRITER'S GOAL

In preparing this book my main goals have been to survey the already-invented schemes for reducing heat-loss through windows, invent schemes that fill the gaps between existing schemes, and propose some radically new schemes.

In describing the various schemes, I have concentrated mainly on the physical principles involved, i.e., on the rationales.

Throughout, emphasis is on devices for use on double-glazed windows. Most persons concerned about heat-loss through windows have already installed storm windows or have windows of Thermopane, Twindow, or the like. Nearly all solar-heated houses have double-glazed windows. Passive solar houses have especially large (south) double-glazed windows. Of course, most of the schemes described here are applicable also to single-glazed windows.

Although many of the shutters and shades described here are well proven and commercially available, many untried schemes are described also. Most of these may prove successful—if the designer and builder display skill in working out the details. Some may prove to be impractical.

## DEFINITIONS OF SHUTTERS, SHADES, ETC.

**Shutter** A rigid insulating plate, or set of closely connected rigid plates or rigid strips, used to cover a window for the purpose of reducing heat-loss. It may be applied to the outdoor side of the window or the indoor side, or between glazing sheets. It may be permanently attached, e.g., by hinges, or may remain unattached so as to be instantly removable. It may be thick or thin and may be opaque, translucent, or transparent. Some kinds of shutters (those that include many rigid parallel strips) can be rolled up.

**Venetian blind** A set of horizontal opaque strips (or slats or vanes) that can be rotated at least 90 degrees about horizontal axes so as to (a) form a contiguous, nearly coplanar set that greatly restricts passage of light and air or conversely (b) form an open set that freely permits such passage.

**Shade** A device that consists mainly of one or more flexible sheets or quilts and can be unrolled (or vertically raised or lowered) to cover a window and, when not needed, can be moved out of the way. The device may be used to provide privacy, reduce heat-loss, or prevent solar radiation from penetrating deep into the room. It may be attached at the top or bottom of the window and may be opaque, translucent, or transparent. The roll-up process may be accomplished by an automatic spring-return or by other means.

Note: A device that can be unrolled or rolled up and consists mainly of rigid strips is called a shutter, as explained above.

**Curtain** A flexible sheet (e.g., of cloth) or a set of several such sheets in series, used to cover a large portion of a window or cover a frame or space beside a window. The device may be movable; if it is, the motions are mainly horizontal. The purpose of the device may be to provide visual privacy, to diffuse daylight that enters the room, to reduce heat-loss, or to make the window region more attractive. Some curtains are translucent; some are opaque. Ordinarily, curtains are drawn aside at the start of the day.

**Quilt** Several flexible sheets of cloth or plastic (or plastic foam) attached to one another in series to form a thick, soft, flexible assembly. Quilts constitute the hearts of some shades and curtains.

**Drape** A curtain, especially a long curtain.

**Screen** A thin object, usually of wire mesh, that excludes flies, mosquitoes, etc., but permits passage of light and air.

**Valance** A fixed ornamental device, usually of fabric, that covers the gap and hardware associated with the uppermost portion of a shade or curtain.

**Canopy** A rigid housing or cover that encloses the hardware (roller, e.g.) associated with the uppermost portion of a shade or curtain.

## IS THERE SOME ONE BEST CHOICE OF SHUTTER OR SHADE?

Clearly, there is no single best choice. Different situations require different solutions. Consider these contrasting situations:

### Location of House

Hot climate (e.g., Florida)/Cold climate (e.g., Maine)

Windless valley/Windy hilltop

### Location of Window in House

First story (outdoor shutter easily reached from ground)/Upper story (outdoor shutter, reached by ladder)

Leeward side of house/Windward side

South side/North side

### Window Size and Shape

Large window; large ratio of area of perimeter/ Small window

Wide window/Slender window

Deeply recessed window/Window not recessed

Elaborately framed window/Simply framed window

Top of window is only 6 ft above floor level/Top of window is much too high to reach

### Window Style

Double-hung sashes/Casement sashes

Wooden sashes/Metal sashes

One large pane/Many small panes

### Window Condition

Double glazed/Single glazed

Airtight/Very leaky

Thick Plexiglass glazing/Thin glass glazing

Existing expensive shade or curtain/No shade or curtain

### Use of Window

Used for view only; is never opened/Used also for ventilation; must be opened often

### Humidity in Room in Winter

Low. Moisture condensation is not likely to occur/ High. Moisture is major problem

### Risk of Fire

Low. None of the occupants smoke. There are no wood stoves/High. Use of flammable materials is ruled out.

### Risks of Damage by Children

Low. No children in household/High. Several rambunctious children in household

### Performance Required of Shutters or Shades

Resident demands high thermal performance, i.e., large saving of heat/Resident satisfied with modest saving of heat

Resident demands high durability/Resident satisfied with moderate durability

Resident willing to spend 15 minutes a day opening and closing shutters and shades/Resident wants system to be fully automatic

Resident willing to assign space for storing removable shutters/Resident unwilling to relinquish such space

### Cost Considerations

Cost is not a major consideration/Only very-low-cost devices can be considered.

### Esthetic Requirements

Resident is indifferent to appearance/Resident greatly concerned about appearance

### Procurement and installation

Shutter or shade is to be purchased/It is to be made by the resident himself

It is to be installed by a professional/It is to be installed by the resident himself

A 3-hour installation time is acceptable/Installation must take no more than 30 minutes

### Maintenance

Resident is willing and able to make minor repairs from time to time/Resident cannot make repairs. The shutter or shade must be maintenance free.

### Additional Benefits Desired

None/Shutter or shade must help exclude burglars, help keep heat out in summer, help collect solar energy in winter.

Clearly, a great variety of designs is needed. Each designer and each buyer should have the above-listed considerations in mind when deciding which kind of shutter or shade to use.

### USE OF A MIXTURE OF STRATEGIES

Sometimes it pays to use a mixture of strategies: one may treat certain windows differently, or even treat portions of a given window differently.

**Different windows**  If there are several windows in a room, one may equip the most important one with an expensive shutter that reduces nighttime heat-loss greatly and is removed each morning; thus there is no sacrifice of daytime view. A less important window (one facing a garage, for example) may be equipped with a translucent shutter that, although highly effective thermally, is very cheap. The device may be left in place night and day; the fact that it blocks daytime view is not important.

An airtight window may be equipped with a low-cost device that provides no tight seals, whereas a very leaky window may preferably be equipped with an expensive device that has tight seals.

In general, each window should be judged on its own merits. By using a mixture of strategies, the designer may get better overall performance at lower overall cost. Of course, in some instances the use of many kinds of devices in one room would produce an odd and unacceptable appearance.

**Portions of a given window**  One could cover the lower half of a window with a device that is to be left in place all winter, night and day, greatly reducing heat-loss and yet costing little. The upper half could be equipped with a cheap device that is removed each morning; alternatively it could be equipped with an expensive transparent device that is left in place night and day.

If the lower half of a window is normally covered by a gauzy curtain, this half could be equipped with a translucent insulating device that is left in place night and day. For the upper half, a very different treatment might be appropriate.

### FUNCTIONS OF A WINDOW

A window can perform many functions, including:

admitting daylight

allowing people inside to view the outdoor scene

allowing outdoor air to enter, or allowing indoor air to leave; i.e., providing ventilation

excluding rain, snow, flies, mosquitoes, etc.

excluding burglars, neighbors' children, squirrels, birds, etc.

preventing heat-loss to outdoors in winter

allowing solar radiation to enter in winter

excluding solar radiation in summer

reducing conduction of heat from outdoors in summer

serving as emergency means of egress—for example, when the building is on fire serving as emergency means of ingress—for example when the doors are locked and you have lost your key—and no policeman is watching you!

## MAIN TYPES OF WINDOWS

Presented below are descriptions of a few of the main types of windows. The descriptions are brief because this is a book about shutters and shades, not windows.

### Ordinary Double-Hung Window

This type of window, in common use in the U. S. for over 100 years, has two sashes: upper and lower. Each can be slid vertically and each is supported at the sides by a cord, metal band, or chain. Attached to the free end of each sash cord is a heavy iron sashweight, serving as a counterweight. Each sash cord runs over a pulley. The sashweights are situated in slender vertical boxes, or channels, within the jambs, i.e. sides, of the fixed frame of the window. When the window is closed, the two horizontal, contacting, sash edges can be pulled snugly against each other by means

of a centrally located, spirally tapered, metallic lock. There is a slender vertical strip of wood (retaining strip, or parting strip) affixed to the right side of the fixed frame of the window and a similar strip affixed to the left side; these strips prevent the upper part of the upper sash from moving inward and prevent the lower part of the lower sash from moving outward. Because the pertinent edges of the lower sash are cut away just enough to accommodate the strips, the two sashes maintain contact with one another as each slides up or down.

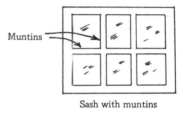

Sash with muntins

Many sashes contain *muntins*, or *mullions*, which divide the window area into several small areas.

Note: On consulting Webster's Third New International Dictionary, Unabridged, one finds that the term *muntin* is preferable to *mullion* for the horizontal and vertical divider strips depicted in the accompanying sketch.

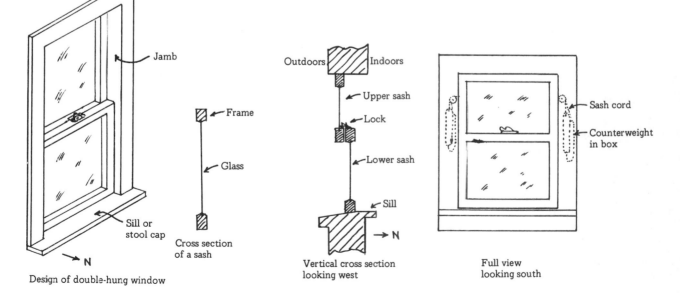

Design of double-hung window

Cross section of a sash

Vertical cross section looking west

Full view looking south

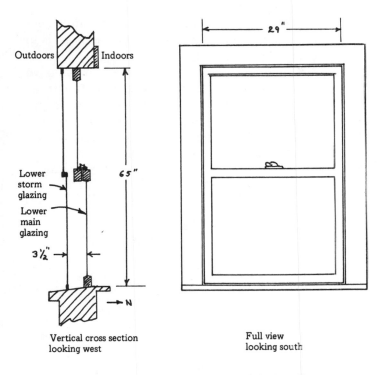

Outdoors    Indoors

Lower
storm
glazing

Lower
main
glazing

3½"

Vertical cross section
looking west

29"

65"

N

Full view
looking south

Double-hung south window of house at 19 Appleton St.,
Cambridge, Mass.

## Casement Window

This type, being hinged at one side, is similar to a door. There are various means of opening and closing such windows; often a crank-and-arm system is provided. Suitable locking devices are included.

## Fixed Window

Some windows are fixed—permanently shut. There are no hinges, counterweights, or locks. The edges can be sealed with caulking material.

## Double (or Double-Glazed) Windows

Many modern windows include two sheets of glass, about 3/16 or 1/4 in. apart, that are joined with a more-or-less airtight seal (to exclude moisture and dust) or may actually be fused (welded) together.

Two glass sheets welded
at edge

Two common brands are Thermopane, made by Libbey-Owens Ford Co., and Twindow, made by PPG Industries, Inc.

In older houses, storm sashes (storm windows) are often used. Such sashes, together with the main sashes, provide two layers of glazing. The thickness of the airspace between them is about 1 to 4 inches.

## Monotask Windows

Prof. R. S. Levine of the University of Kentucky has developed three highly specialized types of windows. One type emphasizes admitting daylight, another emphasizes view, and the third emphasizes ventilation. Used together in proper proportions, they provide many benefits yet cost relatively little. One benefit is reducing heat-loss on winter nights. The three types are described below.

**Windows for admitting daylight**  These windows are situated high above floor level; thus the transmitted daylight penetrates deep into the room. The glazing is translucent; accordingly the daylight spreads to illuminate all parts of the room. Many (cheap) translucent sheets are used in series; thus heat-loss by conduction is very small. The outermost sheet is permanently attached and sealed; thus there is no heat-loss by leakage of air. The windows are made small enough to fit between the studs, which, typically, are 24 inches apart on centers; thus the windows can be built on-site as the house is being built.

**Windows for view**  These windows, also built on-site, are multi-glazed with transparent sheets of glass which are fixed and permanently sealed. Thick indoor shutters are provided and are closed manually on especially cold nights.

**Windows for ventilation**  Some of these windows —actually *ports* in most instances—are at near-ceiling height and some are at near-floor height. When they are open, cold air enters the room via the lower ports and warm air escapes via the upper ones. A high rate of passive ventilation is achieved. In winter the ports are permanently closed by inexpensive, thick, opaque, heavily insulated shutters.

A more complicated window-and-shutter system developed by R. S. Levine is described in Chapter 15.

## CONVENTIONS USED IN DESCRIPTIONS AND DRAWINGS

The descriptions presented in later chapters are meant to emphasize principles: principles of design and layout, principles of opening and closing, principles used in choice of materials. Usually no effort is made to specify exact dimensions, exact choice of materials, or exact methods of fabrication and installation.

The drawings are greatly simplified, not merely to save effort but also to give emphasis to the important features.

The portions of houses shown are of repulsive shape! They are included only to make the drawings more readable, i.e., to help the reader recognize the relationship of the window to the house.

When I show a vertical transverse section of a window, I assume that south is to the left and the observer is looking west. The double-glazing is indicated by two vertical lines.

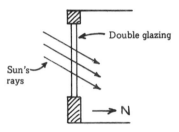

When I show a front, or full, view of the window, I assume the observer is standing in the room and is looking south. (Exception: for *outdoor* devices, the opposite convention is used.)

Transparent or translucent objects are sometimes hatched thus:

Opaque objects are sometimes hatched thus:

Numbering system used: I have assigned numbers to each distinct design of shutter or shade. The numbers are entirely arbitrary and are intended to serve only for quick identification. Examples are "Scheme 8.7," "Scheme 12.5." Minor modifications of Scheme 8.7 may be called "Scheme 8.7a," "Scheme 8.7b," etc.

Various fittings, etc., are portrayed as indicated here.

| Fitting | Perspective view | Front or side view |
|---|---|---|
| Button | | |
| Ear | | |
| Tab (or clip, or spring bracket, or flexible finger) | | |
| Hinge | | |

## WARNINGS

Six important warnings are presented below.

**Credits for inventions**  Some of the schemes described were invented or developed by identifiable persons. In many cases I have included the names of such persons.

Some of the schemes are of unknown origin. Perhaps friends will soon tell me the inventors' names.

Some schemes I invented myself. This is true of many of the schemes for which no inventor is listed. Probably many of these schemes were invented by others, years ago. Many of these schemes are fairly obvious to anyone who puts his mind on the subject.

**Patents**  I myself have not applied for patents on any of the devices described here. Why not? Because many of the devices may prove to be mediocre (may cost too much or wear out too soon), or may be too simple and obvious to patent, or may have been invented long ago by other persons. Also, obtaining patents involves time-consuming

talks with patent attorneys, costs much money, entails delays of many years, and requires keeping mum in the interim.

Some of the schemes invented by other persons have been patented. In most such cases if I know the patent number I state it.

Ordinarily I do not know whether other persons' schemes have been patented and do not know whether patent applications have been filed. I have made no systematic attempt to find out.

Even when the reader is informed that a certain scheme is patented, he may wonder which specific aspects of the scheme are at issue. Also he may wonder whether eventually a court will declare the invention to be trivial and the patent invalid. He may wonder whether—if the patent is an old one—its 17-year life has run out.

**Trademarks**  Many of the materials, components, etc., described in this book have names that are trademarked (or registered).

In general, I have omitted trademark indications, partly to save time and space and partly because, in so many instances, I do not know (a) whether a given term has been trademarked, (b) whether (if trademarked long ago) it has "escaped" into the public domain, or (c) what, exactly, is trademarked ("Widget," or "Widget Shutter," or "Widget high-performance shutter").

My impression is that many pertinent books and periodicals routinely omit trademark indications.

**Accuracy**  Much of the information presented here has not been finally verified. No doubt many errors of omission and commission are present: errors as to design and operation of the devices described, errors as to the inventors or companies involved etc. Good (likewise bad) features of the devices may be overstated or understated. No reliance should be placed on the information presented unless independent verification is made.

**R-Values**  Although manufacturers of shutters and shades usually state the R-values of their products, I usually decline to do so. There are many reasons for this refusal (see Chapter 4).

R-values of materials, for example a 1½-in. sheet of Styrofoam or a ½-in. sheet of Thermax, are meaningful and reliable. But the values that manufacturers may give for specific shutters or shades installed (with unspecified care) on actual windows (of unspecified type) in houses subjected to actual winds (of unspecified speed and direction) may have little meaning or may be misleading.

**Early state of the art**   The development of widely applicable, low-cost thermal shutters and shades is in an early state. The urgent need for such devices has arisen only in the last few years. Important new materials have become available in this same period. Many inventors are now hard at work testing today's shutters and shades and inventing new kinds. Much progress is to be expected.

Accordingly, this book is merely an interim report—not final or definitive. In a few years a very much better book can be written.

**Main Producers of Thermal Shutters and Shades**

| Producer | Product | See page |
|---|---|---|
| American German Industries, Inc. | Rolladen shutter | 85 |
| Appropriate Technology Corp. | Window Quilt Thermal Curtain | 159 |
| Ark-Tic-Seal Systems, Inc. | In-Sider | 167 |
| Boyle Interiors Co. | Reversible Piggy Back | 174 |
|  | Dimout |  |
|  | Blackout |  |
|  | Wind-n-Sun Shield |  |
| Center for Community Technology | Roman Shade II | 179 |
| Conservation Concepts Ltd. | Warm-In | 175 |
| (Cornerstones: see Homesworth Corp.) |  |  |
| Green Mountain Homes, Inc. | Thermo-Shutter | 132 |
| Homesworth Corp. | Sun Saver | 130 |
| Insul Shutter, Inc. | Insul Shutter | 132 |
| Pease Co. | Pease rolling shutter | 87 |
| Plaskolite, Inc. | Weatherizer Kit In-Sider | 106 |
| Rolscreen Co. | Slimshade Rolscreen Pella | 100 |
| Rolsekur Corp. | Rolsekur | 88 |
| Shutters, Inc. | Thermafold | 66 |
| Solar Science Industries | Kool-Shade | 89 |
| Solar Survival | SUNLOC | 98 |
| Technology Development Corp. | Weatherguard | 158 |
| Therma-Roll Corp. | Therma-Roll | 86 |
| Zomeworks Corp. | Beadwall | 101 |
|  | Silli Shutter | 141 |
|  | Roll-Away Magnetic Curtain | 160 |

# CHAPTER 2

# HOW HEAT IS LOST THROUGH WINDOWS

- Typical Temperature Distributions at Windows
- Mechanisms of Energy Loss
- What Is Heat?
- The Electromagnetic Spectrum
- Amount of Power Radiated
- Rate of Heat-Loss Through Window by Radiation
- Rate of Heat-Loss Through Window by Conduction
- Rate of Heat-Loss Through Window by Convection
- Overall Rate of Heat-Loss Through Well-Sealed Window on a Winter Night
- Heat-Loss Through Well-Sealed Window During Winter as a Whole
- Circumstances That May Invalidate Predicted Values of Heat-Loss
- Relative Importance of Heat-Loss Through Windows
- Justification for Disregarding Solar Energy Inflow

This chapter deals with *how* heat flows and *how much* heat flows. Emphasis is on heat-loss through double-glazed windows on winter nights. Background information on these topics is presented in Appendixes 1–3.

11

## TYPICAL TEMPERATURE DISTRIBUTIONS AT WINDOWS

In winter, in the immediate neighborhood of a typical window, the temperature varies smoothly from indoors to outdoors. The distribution, i.e., the graph of temperature vs. position along a line normal to the window, has about the form one would expect: a steady decrease from a location near the center of the room to an outdoor location several inches from the window. The decrease is especially steep at locations very close to a glazing sheet (glass pane).

The first of the accompanying graphs shows the distribution near the east window of the first-story den of my Cambridge, Mass., house. The window has upper and lower main sashes and upper and lower storm sashes. Between the inner and outer lower sashes there is a 3½-in. space. The window is reasonably tight; there are few obvious cracks. Because there are so many trees and houses nearby, the windspeed within 1 ft. of the window is usually less than 2 mph. To find the temperature distribution, I installed a lot of ordinary $2 thermometers along a line normal to the window (see sketch and also see Appendix 1).

The second graph shows the temperature distribution when a high-quality indoor shutter (1½-in. plate of Styrofoam SM) is in use, pressed against the sash.

Note: These graphs are not to scale and not accurate. They are intended merely to show the general shape of the temperature distribution curve before and after an indoor shutter is installed.

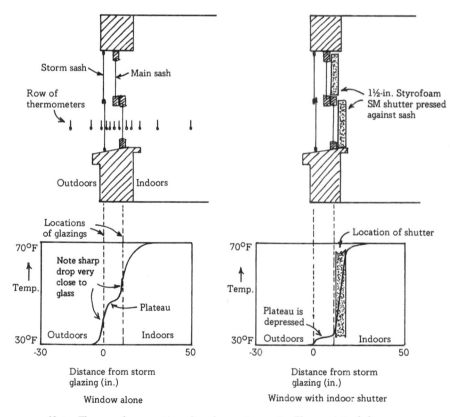

Window alone                  Window with indoor shutter

Note: These graphs are not to scale and are not accurate. They are intended merely to show the general shape of the temperature distribution curve before and after an indoor shutter is installed.

Different kinds of shutters produce different distributions, as suggested by the following graph.

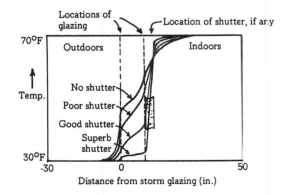

Distance from storm glazing (in.)

Obviously, the shape of the curve depends on many factors. For example, if much cold outdoor air leaks in through the window, the plateau is lowered. If much *warm* indoor air leaks out through the window, the plateau is raised. The between-glazings window-jamb areas may raise or lower the plateau depending on whether these areas are thermally linked mainly to indoors or to outdoors. Outdoor wind and indoor air currents can change the curve shape considerably.

## MECHANISMS OF ENERGY LOSS

On winter nights, energy flows through a window by these three well-known mechanisms: radiation, conduction, and convection. It flows along many different paths, e.g., through the glass panes, through the muntins and frames of the sashes, and through the fixed frame of the window as a whole. Cracks may be important: if there are large cracks at the edges of the panes, at the edges of the sashes, or in the fixed frame, much energy may be lost by out-leak of room air—or by in-leak of cold outdoor air and corresponding out-leak elsewhere, e.g., via the attic.

Several of these losses are hard to calculate, (i.e., hard to predict), and also hard to measure. In some houses, leakage of air through the windows is the main cause of energy-loss; yet there is no satisfactory way of describing the cracks or of predicting the heat-loss. Losses by radiation can often be calculated with fair accuracy and the same applies to losses by conduction. But losses through the fixed frame of the window (especially if there are hollow boxes for sash weights and holes for pulleys and sash-cords) can be estimated only crudely.

## WHAT IS HEAT?

Heat is the energy of motion of atoms and molecules: the kinetic energy of their translational, vibrational, and rotational motions.

It is not a sensation. The heat in a bucket of hot water exists irrespective of whether anyone sticks his hand in and feels the warmth.

It is *not* temperature. A bucket of lukewarm water may contain more heat than a cup of boiling water.

*It is merely energy:* one particular kind of energy, namely kinetic energy of a large quantity of atoms and molecules. It is the most common and least glamorous kind of energy.

In the U. S., the most-used unit of heat is the British thermal unit (Btu).

Should radiation be called heat? No. It is energy, but not heat. A clear-cut distinction exists between radiation and heat: radiation can be reflected by mirrors, focused by lenses, and dispersed by prisms and gratings. Heat cannot.

Some people call infrared radiation heat. But this is a mistake. It is energy, but not heat.

All kinds of radiation, whether in the visual or infrared or radio range, can be converted to heat. No kind of radiation is more "heat-like" than any other kind.

It is entirely correct to say that a certain hot body loses heat, and it is entirely correct to say that one of the ways it loses heat is by emitting radiation. In other words, heat-loss by radiation is a perfectly sound expression. But the radiation is not heat.

Warning: I make a strong point of distinguishing between flow of heat and flow of radiant energy. Most writers do not. They use the word "heat" for both kinds of energy flow—which makes their discussions simpler looking but sometimes wrong.

The painful fact is that, typically, two quite different kinds of flow exist and these may be related or unrelated. In extreme cases the flows may be in different directions! Having worked for over 40 years with many kinds of radiation (gamma, x-rays, UV, visual, near-IR, far-IR, and radio), I am unwilling to shut my eyes to the basic differences among the many forms of energy flow.

## THE ELECTROMAGNETIC SPECTRUM

The accompanying figure, employing a logarithmic scale, shows the electromagnetic spectral range and its main subdivisions. The subdivisions are, of course, arbitrary. Nature herself knows of no subdivisions; the spectrum is one smooth totality. The overall extent of the range is infinite; the graph could continue indefinitely to the left and to the right. The portions of the spectrum of daily importance in science and industry cover more than 50 octaves. (An octave is a range in which the wavelength doubles.)

### Infrared Radiation

Radiation in the wavelength range from 0.7 to 80 microns (or micrometers, or $10^{-6}$ meter) is called infrared (IR). It is of great importance to solar heating engineers and to persons designing thermal shutters and shades.

The infrared range contains two sub-ranges of great (and separate) interest:

| | |
|---|---|
| Near-infrared (near-IR) | 0.7 to 3.0 microns. Solar radiation includes much such radiation. |
| Far-infrared (far-IR) | 3 to 80 microns. All warm objects (actually, all objects except those that are at absolute zero) emit such radiation. The hotter the object, |

the more it emits. Some mirror-like objects (polished aluminum sheets, for example) emit relatively little. Nearly all household objects, including furniture, walls, rugs, and people, emit strongly.

The two curves of the accompanying figure show (1) the solar spectrum and (2) the spectrum of radiation emitted by household objects that are at about 70°F. The latter spectrum has been greatly magnified by the draftsman to facilitate comparison. Again a logarithmic wavelength scale is used.

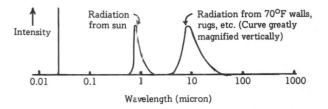

The exciting fact is that these two spectra are separate—distinct. There is no significant overlap. Because they are separate, they can be dealt with separately. For example, a chemist can make a plastic film that absorbs in one of these spectral ranges but not the other or reflects in one but not the other, or transmits in one but not the other. Persons developing selective absorbers, selective reflectors, or selective transmitters take advantage of the fact that the two spectra are separate.

Whereas solar heating engineers are interested mainly in the near-IR, insulation engineers and designers of thermal shutters and shades are interested mainly in the far-IR.

The accompanying graph compares the far-IR spectra of typical objects at temperatures of 70°F

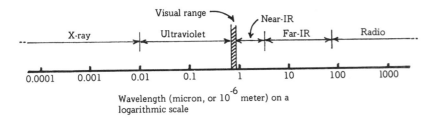

and 30°F. In each case the object is supposed to have the maximum possible emissivity (black-body-type emissivity). In each case most of the energy is in the 6-to-20 micron range and practically all of it is in the 3-to-80 micron range.

The peaks of the curves for the 70°F and 30°F objects are at 9.9 and 10.7 microns respectively. Roughly speaking, the curves have the same shape and almost the same peak wavelength (about 10 microns). In each case about 25% of the energy is at wavelengths *shorter* than the peak wavelength, and about 75% is at wavelengths *longer* than the peak wavelength.

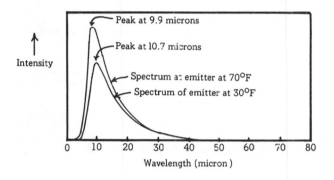

The hotter an object is, the shorter the wavelength of the peak of the curve. Specifically, the wavelength of the peak varies inversely with the absolute temperature of the body in accordance with Wien's displacement law. If wavelength is expressed in microns and absolute temperature is expressed in degrees Kelvin, the exact formula is:

$$\text{Peak wavelength} = \frac{(2.9 \times 10^3 \text{ micron } °K)}{T}$$

and if temperature is expressed in degrees Rankine, the formula is:

$$\text{Peak wavelength} = \frac{(5.22 \times 10^3 \text{ micron } °R)}{T}$$

70°F corresponds to about 530 on an absolute scale (Rankine scale, which starts at −459.67°F). 30°F corresponds to about 490 degrees Rankine. Thus the peak wavelengths for emitters that are at 70°F and 30°F are 9.9 micron and 10.7 micron, as

stated above. (Notice that the two absolute temperatures differ by about 7.5% and the two peak wavelengths differ by about this same amount.)

## AMOUNT OF POWER RADIATED

The hotter an object is, the more power it radiates. The amount varies with the fourth power of the absolute temperature in accordance with the Stefan-Boltzmann law:

Power = $(1.71 \times 10^{-3})$
    $\times$ (absolute temperature in degrees Rankine)$^4$.

The accompanying table lists the amounts of power radiated $(Btu/(ft^2 hr))$ by ideal-blackbody emitters $(\epsilon = 1)$, and also by bodies for which $\epsilon = 0.5$ and 0.1, assuming various temperatures of the emitting body.

**Power Radiated by Unit Area of a Body That Has Emittance 1.0 or 0.5 or 0.1**

| Temperature of Body | Power radiated $(Btu/(ft^2 hr))$ | | |
|---|---|---|---|
| °F | When $\epsilon = 1$ | When $\epsilon = 0.5$ | When $\epsilon = 0.1$ |
| −40 | 53 | 27 | 5.3 |
| 0 | 77 | 38 | 7.7 |
| 30 | 99 | 49 | 9.9 |
| 32 | 100 | 50 | 10 |
| 50 | 115.7 | 58 | 12 |
| 68 | 133 | 67 | 13 |
| 70 | 135.0 | 68 | 14 |
| 86 | 152 | 76 | 15 |
| 90 | 157 | 78 | 16 |
| 100 | 168 | 84 | 17 |
| 104 | 173 | 87 | 17 |

A good benchmark figure to remember is: A one-square-foot blackbody surface at 70°F emits 135 Btu per hour.

Notice this is about half the amount of power received (by a one-square-foot surface) from the sun at noon on a fairly sunny day. How can this be true? How can a black body at room temperature give out half as much radiation as a similar area receives from the blazing hot sun? The answer is that the sun's radiation arrives, mainly, within a very slender range of directions—within a ½-deg. cone—whereas the radiation emitted by the black body spreads out in all directions. If one were to ask: "How much power does the 70°F blackbody radiate *just within a ½-deg. cone?*," the answer would be: "Many thousand-fold less!."

## RATE OF HEAT-LOSS THROUGH WINDOW BY RADIATION

Various preliminaries have been completed. The nature of radiation has been discussed. The shapes of the pertinent spectra have been displayed. The amounts of radiation emitted by warm objects have been tabulated.

What does all this mean in terms of heat-loss by passage of far-IR radiation through a window? Let us consider three especially interesting cases:

**Case 1:**  The glazing transmits far-IR radiation freely. This applies, for example, if the window is glazed just with sheets of very thin plastic. Inasmuch as indoor objects (walls, chairs, etc.) are warm and outdoor objects (ground, trees, air, etc.) are cold, much radiation passes outward through each square foot of the thin plastic and a somewhat smaller amount passes inward. The difference is the net heat-loss by radiation. The difference can be fairly large.

**Example:**  Consider a window glazed just with a 0.001-in. film of polyethylene. Suppose that the indoor and outdoor temperatures are 70°F and 30°F and that the indoor and outdoor objects have emittances of about 1. What is the heat-loss by radiation? The outward and inward flows are about 135 and 99 Btu/(ft² hr) respectively and the difference is about 36 Btu/(ft² hr). This is the net heat-loss-by-radiation through the plastic film.

I say no more about this case because it is so rare. Seldom does one find a window that has high transmittance for far-IR radiation. Such a window would lose a ridiculously large amount of energy by radiation each winter.

**Case 2:**  The glazing completely reflects far-IR radiation. (No actual glazing has this extreme capability. But some specially coated glazing sheets come close.) Much far-IR radiation from indoors strikes the glazing and a somewhat smaller amount from outdoors strikes it—but all such radiation is immediately reflected! It is sent "back toward where it came from." Thus there is no heat-loss by radiation.

True, the glazing may transmit much heat by conduction. However, none is transmitted by radiation.

**Case 3:**  The glazing absorbs practically all of the far-IR radiation incident on it. This applies when the window is made of glass, provided that the glass is at least 1/16 in. thick. (Most sheets of glass are thicker than this.)

Here there is no direct, one-step loss of energy (from indoors to outdoors) by radiation.

**Example:**  Consider a window that is single-glazed with ⅛-in. glass. Suppose that the indoor and outdoor temperatures are 70°F and 30°F. What is the one-step, indoor to outdoor heat-loss by radiation? Answer: None. The glass intercepts and absorbs practically all of the far-IR radiation incident on it.

Note: Although there is no simple direct heat-loss by radiation, there is an important indirect effect. Much far-IR radiation travels from room objects to the glass and is absorbed by it. This warms the glass, which thereupon loses much heat by conduction. Also the glass itself emits far-IR radiation; about half of this goes inward, into the room, and half goes outward, to the outdoors. Thus far-IR radiation plays a role even though there is no simple direct heat-loss by radiation. The subject is too complicated to discuss in detail here.

What about the flow of energy by radiation traveling through air? Consider a ½-in. airfilm on the indoor side of a window; or a 10-ft-long region of air on the indoor side of a window; or a ½-in. or 10-ft region of air on the outdoor side. How freely does far-IR radiation travel through such air?

The answer is that it flows almost infinitely easily and at the speed of light. The flow is virtually the same as if there were a vacuum. However one may choose to define the resistance to this flow, the value of the resistance is zero.

(There is much resistance to heat-flow-through-air by conduction and convection. This is discussed in a later section.)

## RATE OF HEAT-LOSS THROUGH WINDOW BY CONDUCTION

Heat (i.e., energy of molecular motion) flows by conduction through every kind of material, whether solid, liquid, or gas. (A vacuum can stop it—but to maintain a vacuum despite the tremendous pressure of the atmosphere (14.7 lb/in.$^2$) is very costly if the area involved is large.)

Metals conduct heat readily, glass less readily, plastics still less readily, and various foam-like or fibrous materials (polystyrene foam, isocyanurate foam, urethane foam, fiberglass, etc.) conduct it very poorly.

Note: The discussion here is limited to a single method of transfer of energy, namely conduction. No account is taken of transmission of energy by radiation and transmission of energy by convection. These methods are discussed elsewhere.

The ability of a material (for example, Styrofoam) to conduct heat is called the material's *conductivity k*. It is defined as the amount of heat, in Btu, passing (in one hour) through a plate that is 1 in. thick and one square foot in area when one face of the plate is 1 F degree hotter than the other face.

Often the reciprocal quantity—the *conductive resistance-per-inch*—is used. It is defined as 1/k.

### Definitions of C, U, R

Often an engineer is interested in the properties of a plate as a whole—a plate that has a thickness very different from 1 inch. The ability of such a plate to conduct heat is called the *thermal conductance C*. The reciprocal quantity is called the *thermal resistance R*. Often, the symbol "U" is used instead of "C." The difference, according to the *1977 ASHRAE Handbook of Fundamentals*, is that C is supposed to refer just to the plate proper whereas U is supposed to refer to the plate plus flanking air-films. U has many names, including thermal trans-

Illustrative Approximate Values of Conductivity k and Resistance-per-Inch

| Material | k Btu in. / ft$^2$ hr°F | 1/k ft$^2$ hr°F / Btu in. |
|---|---|---|
| Aluminum | 1400 | 0.0007 |
| Copper | 2500 | 0.0004 |
| Glass | 10 | 0.1 |
| Lucite or Plexiglas | 2 | 0.5 |
| Oak wood | 1 | 0.9 |
| Pine wood | 0.8 | 1.25 |
| Isocyanurate foam | 0.13 | 8 |
| Urethane foam | 0.14 | 7 |
| Polystyrene foam | 0.2 | 5 |
| Fiberglass | 0.28 to 0.33 | 3 to 3.5 |
| Brick | 5 to 10 | 0.1 to 0.2 |
| Concrete | 9 | 0.12 |

Sources: Various.
Note: Values do not take into account the flanking air-films.

mittance, overall coefficient of heat transfer, and overall coefficient of heat transmission. Unfortunately many writers have failed to understand the distinction between C and U, with the result that their calculations may be incorrect.

The unit of C and U is: Btu/(ft°F hr$^2$).

The reciprocal of C or U is called R, the resistance, or thermal resistance, or total thermal resistance. Its unit is: (ft$^2$ hr °F)/Btu. Various subscripts may be used. $R_r$ is the resistance associated with energy flow by radiation. $R_{cv}$ is the resistance associated with heat flow by the important thermal convection process. $R_t$ is the total resistance associated with energy flows in which both radiation and thermal convection are important.

### Application of the Quantity k

Using data such as are tabulated above, anyone can quickly calculate how much heat will be conducted through a 1-in.-thick opaque plate of any given material when the temperatures of the faces of the plate are known. The amount is simply the product of (1) k, (2) the area of the plate in square feet, (3) the temperature gradient in F degrees per inch, and (4) the period of time in hours.

**Example:** How much heat is conducted in 12 hours through a 6 ft$^2$, 1-in.-thick urethane foam panel when the faces are at 70°F and 30°F?

The answer is:

$$[(0.14 \text{ Btu in.})/(ft^2 \text{ hr } °F)][6 \text{ ft}^2]$$
$$\times [40°F/\text{in.}][12 \text{ hr}] = 403 \text{ Btu.}$$

Suppose that one has a plate that is not 1 inch thick. To find the amount of heat it transmits by conduction, one must divide the value computed in the above-described manner by the thickness in inches. For example, if the above-specified urethane plate were 4 inches thick (and the temperature difference between the two faces were still 40°F), one must divide the result presented above by 4 because the gradient is only ¼ as great. Thus the heat-loss is 403/4 = 101 Btu.

Unfortunately, one usually does *not* know the temperatures of the faces of the plate. One knows only the temperature of the air some distance away (e.g., 1 in. away, or 10 ft away) on each side. This brings up the complicated subject of conduction of heat through air.

**Conduction of heat through air** Curiously enough, it is virtually impossible to define the pure process of thermal conduction of heat through air. Heat *is* conducted, but energy is transmitted also by radiation—and the radiation travels much faster. Thus the process of thermal conduction may be masked by the process of radiation. Can one arrange some laboratory apparatus that will permit thermal conduction but will exclude all radiation? No. The air itself is constantly producing (emitting) radiation. Also, every container ever made by man has at least a little emissivity (even polished silver does). If one had a very special box that somehow contained no radiation, it would become "filled" with radiation spontaneously within a fraction of a second. In summary, there is no such concept, even as a thought experiment, as pure thermal conduction of heat through air. (However, if one has a pair of silvered plates that are very close together, the radiant flow is small and the conductive flow is large. Here one can arrive at a

rough estimate of the conductance attributable to the thermal conduction process itself.)

Usually one is forced to deal with the combined process: heat flow by conduction and convection proceeding hand-in-hand with energy flow by far-IR radiation.

Often, long range convection also occurs. This greatly complicates the situation. In writing the following paragraphs I have assumed that long range convection has been prevented.

### Transmission of Heat and Radiation Through Thin Films of Air

Appendix 3 deals with the combined processes of conduction, convection, and radiation in air. Reasonably reliable formulas have been found, partly by using the laws of physics and partly by experimentation. Engineers have arrived at quantities (or numbers) that fit in well with the quantities U and R applicable to an opaque plate. They have arrived at these conclusions:

**Indoor vertical airfilm** Consider indoor air that is within ¼ or ½ in. of a flat vertical sheet. This air is not in turbulent motion—it is nearly stationary (but may be in slow vertical motion). If the vertical sheet is of glass or rigid foam or other material that has a far-IR emittance of about 0.9 or more, the effective R-value of the airfilm in question is about 0.68. (If the sheet has very low emittance, the R-value is greater. For example, if the emittance is 0.20, R is 1.35. If the emittance is 0.05, R is 1.70.) The corresponding conductance is called $h_i$, the indoor combined surface coefficient, according to Yellott (Ref. I-402f, p. 95). When R = 0.68, $h_i$ = 1.47.

**Outdoor vertical airfilm** Consider outdoor air and assume that, in the general neighborhood of the window in question, there is a 12 or 15 mph wind. Outdoor air that is about ¼ or ½ in. from the window is moderately turbulent. Only air that is much closer to the window is nearly stationary. One can assign to this very thin region of nearly stationary air an R-value of 0.17. The reciprocal, called $h_o$, the outdoor combined surface coefficient, is 6.0.

**Combined effect of indoor and outdoor airfilms**

To obtain the combined effect, one merely adds the two pertinent resistances. Thus if one deals with a single, vertical, thin sheet of glass, the total resistance of the system is:

$$R_t = 1/h_i + 1/h_o = 1/1.47 + 1/6$$
$$= 0.68 + 0.17 = 0.85.$$

Actually the glass itself reduces the heat-loss slightly. The accepted (ASHRAE) values pertinent to a vertical sheet of glass of typical thickness and indoor and outdoor flanking airfilms (with outdoor windspeed of 15 mph) are:

$$R_t: \quad 0.91 \ (\text{ft}^2\text{hr} \ °F)/\text{Btu}$$
$$U_t: \quad 1.10 \ \text{Btu}/(\text{ft}^2 \ \text{hr} \ °F).$$

Indoor or outdoor air that is situated beyond the ¼ or ½ in. ranges discussed above is usually moving in various directions, including directions not parallel to the glazing. Thus the air currents may carry much heat toward or away from the glazing and the stationary airfilms flanking it. If one were to try to assign an R-value to air involved in such convective transport of heat, the value would be close to zero.

See Appendix 3 for information on heat-flow from one flat vertical sheet to another with a thin region of air intervening. See also the ASHRAE 1977 *Handbook of Fundamentals*.

**Transmission of Heat and Radiation Through the Combination of an Opaque Plate and Its Flanking Films of Still Air**

To find how much energy is transmitted by the combination of an opaque plate and its flanking indoor and outdoor airfilms, one adds the three pertinent $R_c$ or R values, finds the reciprocal of the sum, and multiplies this by the difference between the nominal indoor and outdoor temperatures.

**Example:** Suppose a 1-in. Styrofoam plate is flanked by indoor and outdoor air, and the nominal indoor and outdoor temperatures are 70°F and 30°F. How much energy flows through one square foot of this system in one hour?

**Answer:** Obtain the sum of the R-values:

| | |
|---|---|
| indoor airfilm | 0.68 |
| 1-in. Styrofoam | 5 |
| outdoor airfilm | 0.17 |
| | 5.85 |

and find the reciprocal: $U = 1/5.85 = 0.17$. Then the energy flow is:

$$(70 - 30) \times (0.17) = (40) \times (0.17)$$
$$= \text{about } 7 \ \text{Btu}/(\text{ft}^2 \ \text{hr}).$$

**Transmission of Energy Through a Thin Aluminum Foil Flanked on Both Sides by Still Air**

Here two airfilms are involved and each is of indoor type. If the aluminum foil is very clean, its emissivity may be 0.05, and in this case the resistance of each airfilm is: $1/h_i = 1.70$. The total resistance is $2(1.70) = 3.40$, almost greater than that of a 1/2-inch-thick plate of Styrofoam. If the foil has become somewhat tarnished or dirty and has an emittance of 0.20, $1/h_i = 1.35$ and the total resistance of the system is 2.7.

**General Case of Transmission of Heat and Radiation Through a System**

To find the amount of energy transmitted by conduction, convection, and radiation, one tries to divide the path into segments (indoor airfilm, first plate, next airfilm, second plate, next airfilm, etc.) and tries to find the effective resistance of each. In some situations this approach fails: there are strange cross-links, strange bypasses. A more complicated analysis is needed.

## HEAT-LOSS BY GROSS CONVECTION

Heat-loss by gross convection is a subject that makes engineers blush. There are complications within complications. Writers tend to deal with this subject last, hoping that the reader will by then be so drowsy that the breadth and depth of ignorance displayed will go unnoticed.

Heat-loss by convection depends on the speed, direction, and turbulence of airflow. It depends on the indoor air current, on the outdoor winds, and on cracks in the window as a whole.

**Indoor convection** Air movements within a house depend on many factors. Is there a forced hot-air system? Does it periodically set the indoor air in violent motion? Is the house a tall one, and does cold air cascade down the staircases? Are there strong drafts near cold walls or fireplaces? Are bedroom windows or basement bulkheads left open?

Such indoor air currents influence the rate of heat-loss through windows, whether well sealed or poorly sealed. Yet to measure the currents is difficult and to estimate how much they affect heat-loss is equally difficult.

**Outdoor convection** Outdoor air movements vary enormously. Most writers assume that on a typical night in winter the wind speed is 15 mph. (Some assume 12 mph.) And they assume that the wind strikes the windows and increases the heat-loss through them—even if they are perfectly sealed.

But are these assumptions realistic? Most houses are in cities or suburbs, and most suburbs are blessed with thousands of trees, shrubs, and hedges. Does the 15 mph speed apply 1000 ft above the houses? Or at 6 ft above the ground? Does it apply 100 ft away from the windows or 1 in. away? Does it apply to one side of the house, two sides, or all sides? I have found that, 1 ft from the ground floor east windows of my suburban house, the air is almost perfectly stationary on winter nights. Seldom does the speed here exceed 2 mph.

Suppose that the windows are deeply recessed and that, close in front of them, there are some tall shrubs. Is the effect of the wind cut in half?

Certainly many houses are on bare plains. Some are on hilltops. In such locations the typical wind speed in winter may exceed 15 mph.

It seems to me absurd for engineers to assume routinely that the typical wind speed in winter is 15 mph with no thought as to whether such value is far too high or far too low for the location in question, with no explanation as to where the windspeed is measured (what altitude, what distance from the window) and no statement as to the direction of the wind in relationship to the orientation of the window in question.

**Leakage of air through cracks in window structures** An especially frustrating topic involving convection is cracks and gaps in window structures—cracks around the panes, around the sashes, and in the fixed frame. The engineering handbooks ask you to measure the lengths and widths of cracks as the first step in computing the amount of air that leaks through the windows. But their request is absurd! Cracks are often inscrutable. Often you can't tell whether they are shallow or extend all the way through to outdoors. You can't arrive at a typical width. The effective length may be hard to judge. Moreover, cracks in a wooden window frame may become wider or narrower as the ambient humidity decreases or increases, causing the wood to shrink or swell.

Suppose you know the widths and lengths of cracks? What then? To find how much air leaks through them, you must know the outdoor-vs-indoor pressure difference caused (1) by wind or (2) by "stack effect," i.e., the tendency for hot air to escape via upper story openings and cold air to enter via lower story openings.

If, by some miracle, you succeed in finding how much cold outdoor air is leaking into the house via cracks, you can easily find the resulting heat-loss. The loss is merely the product of the mass of in-leaking air times the specific heat of air times the indoor-vs-outdoor temperature difference. One cubic foot of 70°F air at sea level weighs 0.077 lb. Its specific heat is 0.24 Btu/(lb.°F). Thus if a cubic foot of 70°F air is replaced by an equal amount of 30°F air, the amount of heat lost is:

$$0.077 \times 0.24 \times 40 = 0.74 \text{ Btu.}$$

With great misgivings, I include a table that presents rule-of-thumb data, drastically rounded off, on heat-loss by leakage of air through cracks in windows and window frames and casings. The data appear in many books, handbooks, etc., but represent, I suspect, little more than a conventionalized set of guesses.

To obtain the last-column data from the next-to-last-column data, one makes use of the fact that one cubic foot of air weighs 0.077 lb. and has a specific heat of 0.24 Btu/(lb., °F). Thus if the indoor temperature exceeds the outdoor temperature by 40°F, the heat-loss associated with 1 ft$^3$ of air is: (0.077

| Fit of window | Is window weather-stripped? | Leakage rate (cubic foot of air per hour) | | Heat loss per hour through cracks of entire 3 ft × 4 ft window when indoor and outdoor temperatures are 70°F and 30°F respectively |
|---|---|---|---|---|
| | | per linear foot of crack | per entire 3 ft × 4 ft window | |
| Poor | No | 200 | 2400 | 1800 Btu |
| Average | No | 100 | 1200 | 900 Btu |
| Average | Yes | 40 | 480 | 350 Btu |
| Excellent | Yes | 10 | 120 | 90 Btu |

lb.)(0.24 Btu/(lb., °F)(40°F) = 0.74 Btu. A leakage rate of 2400 ft³/hr (by a 3 ft × 4 ft poorly fitting, non-weather-stripped window) with indoor temperature exceeding outdoor temperature by 40°F implies a heat-loss rate of (2400 ft³/hr)(0.74 Btu/ft.³) = 1800 Btu/hr.

Comparison: By way of comparison, consider a 3 ft x 4 ft window that is double glazed and has no cracks. Assume that the indoor temperature exceeds the outdoor temperature by 40°F and assume that the windspeed is 12 mph. Then the heat-loss is 275Btu/hr.

A general conclusion is that unless a window is at least fairly well sealed, heat-loss by air leakage is likely to exceed heat-loss by conduction. Once again the moral seems to be: **seal the windows well!**

## RATE OF HEAT-LOSS THROUGH WELL-SEALED WINDOW ON A WINTER NIGHT

The accompanying table presents some approximate values of rate of heat-loss through well-sealed single-glazed and double-glazed windows on a winter night, with various outdoor temperatures and wind speeds. The windows are glazed with glass that is at least 1/16 in. thick.

**Example of use of table**  How much heat is lost in 12 hours through a 10-ft² double-glazed window when the indoor temperature exceeds the outdoor temperature by 50°F and the wind speed is 12 mph? The pertinent number in the table is the one in the lower right corner: 29. The answer is com-

**Rate of Heat-Loss Through Well-Sealed Window on Winter Night (Btu/ft² hr))**

| Temperature difference, indoor vs. outdoor (°F) | wind speed | | |
|---|---|---|---|
| | 2 mph | 6 mph | 12 mph |
| Single-glazed window | | | |
| 1 (U-value) | 0.67 | 1.00 | 1.14 |
| 25 | 17 | 25 | 28.5 |
| 50 | 34 | 50 | 58 |
| Double-glazed window | | | |
| 1 (U-value) | 0.43 | 0.54 | 0.57 |
| 25 | 11 | 13.5 | 14 |
| 50 | 22 | 27 | 29 |

Source: Report by Berman et al. (B-245).

puted thus:

$$(12 \text{ hr})(10 \text{ ft}^2)(29 \text{ Btu}/(\text{ft}^2 \text{ hr})) = 3480 \text{ Btu.}$$

**Note concerning U-values and R-values**  The numbers in the first line of the first half of the table are the U-values of a single-glazed window. The corresponding R-values are the reciprocals of these numbers thus the R-values are 1.5, 1.00, and 0.88. Likewise the numbers in the first line of the second half of the table are the U-values of a double-glazed window, and the reciprocals (2.3, 1.9, and 1.8) are the R-values.

*Note concerning double-glazed windows with very thin intervening air space:* If the air space is only 0.25 in. thick, the U-value is about 20% greater than indicated above for a ½-in. air space. If the air space is only 0.19 in. thick, the U-value is about 25% greater than for the ½-in. airspace. For further details see the ASHRAE 1977 *Handbook of Fundamentals*, p. 22.24.

| Some numbers worth remembering (for 12 mph wind and no in-leak or out-leak of air) | | |
|---|---|---|
| | U | R |
| Single-glazed | 1.14 | 0.88 |
| Double-glazed with ½-in. airspace | 0.57 | 1.8 |

## HEAT-LOSS THROUGH WELL-SEALED WINDOW DURING WINTER AS A WHOLE

To find the heat-loss through a well-sealed window for the winter as a whole, one multiplies the pertinent U-value by 24 (to convert from loss during one hour to loss during one day) and then by the number of degree days pertinent to the location of the house in question.

**Example:** Consider a well-sealed 3 ft × 4 ft, double-glazed window in a location that is near Boston and, typically, has 12 mph winds. What is the heat-loss through this window during a typical winter?

The pertinent U-value (from previous table) is 0.57. The degree-day value is 5500. Therefore the heat-loss through one square foot of the window, per winter, is:

$$(24 \text{ hr/day})(0.57 \text{ Btu/(ft}^2 \text{ hr °F)})$$
$$\times (5500 \text{ deg.-day/winter}) = 75,000 \text{ Btu/(ft}^2 \text{ winter})$$

The loss through the entire window is (12 ft²) times this:

$$12 \text{ ft}^2 \times 75,000 \text{ Btu/(ft}^2 \text{ winter})$$
$$= 900,000 \text{ Btu/winter.}$$

Such a calculation takes no account of heat gains—such as the gains by south-facing windows during sunny hours. In other words, the calculated value presented here is the gross heat-loss. The net heat-loss is smaller.

Often, a designer may wish to know how much heat is lost during the winter nights, or how much heat is lost in the periods from one hour before sunset to one hour after sunrise. An approximate answer may be obtained by taking such a value as is presented above and multiplying it by 0.7 or 0.8. (If the daytime and nighttime were of equal length and the daytime and nighttime temperatures were the same, a factor of about 0.5 would be appropriate. But because the nights are longer and colder than the days, a larger factor should be used. The exact factor depends on the latitude and several other quantities.)

The accompanying table shows the values pertinent to certain locations and three wind speeds.

**Heat-Loss Through One-Square-Foot Area of Well-Sealed Window During Winter as a Whole** (Btu/(ft² winter))

| Location (and degree day figure) | Wind speed | | |
|---|---|---|---|
| | 2 mph | 6 mph | 12 mph |
| *Single-Glazed* | | | |
| Los Angeles (2000) | 32,000 | 48,000 | 55,000 |
| Washington, D.C. (4000) | 64,000 | 96,000 | 110,000 |
| Boston (5500) | 88,000 | 130,000 | 150,000 |
| Pittsburgh (6000) | 96,000 | 144,000 | 165,000 |
| Burlington (8000) | 130,000 | 190,000 | 220,000 |
| Duluth (10,000) | 160,000 | 240,000 | 270,000 |
| *Double-Glazed* | | | |
| Los Angeles (2000) | 21,000 | 26,000 | 27,000 |
| Washington, D.C. (4000) | 41,000 | 52,000 | 55,000 |
| Boston (5500) | 57,000 | 71,000 | 75,000 |
| Pittsburgh (6000) | 62,000 | 78,000 | 82,000 |
| Burlington (8000) | 82,000 | 104,000 | 110,000 |
| Duluth (10,000) | 103,000 | 130,000 | 137,000 |

Note: No account is taken here of energy gains during sunny hours.

## CIRCUMSTANCES THAT MAY INVALIDATE PREDICTED VALUES OF HEAT-LOSS

The heat-loss values presented above must be viewed with skepticism. Actual conditions may differ in so many ways from the assumed conditions! For example:

The indoor temperature may be different from the assumed value of 70°F.

Indoor air currents may play a larger or smaller role than assumed.

The actual degree-day figure may be different from the assumed figure. Locations only 20 miles apart may have degree-day figures that differ by 10%. Also, degree-day figures vary from year to year.

The direction and speed of the wind may be especially favorable or unfavorable. Likewise the orientations of the windows may be especially favorable or unfavorable.

The effect of wind may be increased or decreased by the number and location of trees, shrubs, etc., and also by the extent of recessing of the windows. Height and width of the windows can have an effect also.

The extent of leakage of air through or around the window structure may be different from the assumed extent.

The detailed optical and thermal properties of the glazing may be different from the properties assumed.

Local nighttime sky conditions may be such that the amount of far-IR radiation (from the sky) that reaches the windows is different from the assumed amount.

The solar energy entering the rooms via the windows may be incorrectly included or excluded from the heat-loss calculation.

## RELATIVE IMPORTANCE OF HEAT-LOSS THROUGH WINDOWS

How large is heat-loss through windows relative to heat-loss through ceilings, walls, etc., and losses via air-leakage? No general answer can be given. Circumstances vary so widely!

### Specific Case Considered by Berman et al.

An interesting and possibly typical case is considered by Berman et al., (B-245). They deal with a single-family house with single story, basement, attic, two external doors, and 150 ft$^2$ of window area. Ceiling area is 1300 ft$^2$, and wall area 1000 ft$^2$. Old air is exchanged for fresh air once per hour.

The heat losses (in millions of Btu per 100 degree-days) of this house are tabulated below. Also, percentage-of-total values are included in parentheses. (Warning: I have made some adjustments to the data and some unexplained inconsistencies occur. The amount of heat lost via the basement is assumed to be negligible.)

Sample conclusions:

In an uninsulated house, heat-loss from air exchange does not loom large; but when the house is well insulated and the windows are double glazed, the same amount of heat-loss from air-exchange is relatively large (about 45% of the total).

In a house that has a modest area of windows heat-loss through windows looms large if the

**Heat-Loss (Mbtu per 100 deg.-days)**

| Loss via | No insulation in ceilings or walls; windows single glazed | 4 in. fiberglass in ceilings, 2 in. fiberglass in walls, windows single glazed | same, but with windows double glazed (rough estimate only) |
|---|---|---|---|
| Ceilings | 8.7  (33%) | 1.5  (12%) | 1.5  (14%) |
| Walls | 7.7  (29%) | 2.1  (17%) | 2.1  (20%) |
| Windows← | 4.1  (16%)← | 3.8  (30%)← | 1.8  (17%)← |
| Doors | 0.6   (2%) | 0.4   (3%) | 0.4   (4%) |
| Air exchange | 5.3  (20%) | 4.8  (38%) | 4.8  (45%) |
|  | 26.4 (100%) | 12.6 (100%) | 10.6 (100%) |

windows are single glazed but not if they are double glazed (30% vs. 17%).

Heat-loss through double-glazed windows is not serious in houses in moderate climates, and is not very serious in cold climates unless the aggregate area of windows is large. (In passively solar heated houses, the area is usually very large and the heat-loss may be correspondingly large.)

## JUSTIFICATION FOR DISREGARDING SOLAR ENERGY INFLOW

So far I have taken little or no account of energy inflow via windows. I have concentrated on outflow and disregarded solar radiation that, during daylight hours, enters via the windows and helps warm the rooms.

Actually, solar energy inflow, although valuable, is not very relevant to our present purpose. Even if the inflow through a given window were to exceed the outflow, the fact remains that nighttime outflow is large and could be reduced by use of a shutter or shade. Reducing the outflow at night does not reduce the inflow during the daytime!

If, however, a homeowner contemplates installing opaque shutters or shades and *leaving them on the windows night and day*, an interesting economics question arises: Overall, do such devices do more good than harm? The answer can be yes or no. The issue is a complicated one and is outside the scope of this book.

# CHAPTER 3

# ECONOMICS OF SHUTTERS AND SHADES

- Dollar Value of Heat Lost Through Windows
- Dollar Value of Heat Saved by Shutters
- Maximum Acceptable Cash Outlay for Shutters
- Some Graphs of Heat Saving
- Incremental Cost vs. Incremental Saving
- List of Miscellaneous Penalties and Benefits Associated with Shutters
- Does the Availability of Free Heat Affect the Economics?

Some kinds of shutters and shades cost too much. Various well-made, durable, attractive devices cost $4 to $8 per square foot F.O.B. and, if installed by a professional, may finally cost $5 to $15 per square foot.

Others are cheaper. Some cost almost nothing (see, for example, pages 93 and 115). Some have an F.O.B. cost of $2 to $4; installation may add only 50%, or nothing if the homeowner himself does the work.

What cost is excessive? What cost is acceptable? To answer these questions, one must know the dollar value of the heat lost through the given windows when they are not equipped with shutters or shades.

## DOLLAR VALUE OF HEAT LOST THROUGH WINDOWS

One gallon of oil, weighing 7.65 lb, can produce 7.65 × 18,000 = 138,000 Btu theoretically, or 92,000 Btu actually, if burned in a 67% efficient heating system. Delivery of 1 million Btu (MBtu) requires burning 10.7 gallons. If oil costs $1/gal, the cost of 1 MBtu of delivered heat is $10.70.

As explained on a previous page, one square foot of a well-sealed vertical double-glazed window (of simple, thin glass, with sealed ½-in. airspace between) loses 0.57 Btu per (hr, °F, deg. of temperature difference). Multiplying by 24, one finds the loss per day to be 13.7 Btu per (day, °F, deg. of temperature difference).

To find the heat-loss per winter in a 5500-degree-day location, one multiplies this result by 5500 and gets: 75,000 Btu per ft² per winter.

Multiplying this by $10.7 per delivered MBtu, one obtains 81¢. This is the value, relative to $1/gal. oil burned in a 67% efficient heating system, of the heat lost per winter, at this location (e.g., Boston) by 1 ft² of a well-sealed double-glazed window.*

The corresponding value for a complete, 3 ft × 3.3 ft (area = 10 ft²), well-sealed, double-glazed window in this location is $8.10.

Note: If auxiliary heat is supplied by electrical heaters, the value in question will be higher—say $10 to $25, depending on the local price of electricity.

## DOLLAR VALUE OF HEAT SAVED BY SHUTTERS

Consider a thermal shutter or shade that, when in place on a well-sealed double-glazed window during a night in winter, cuts the heat-loss through this window by 50%. Let us call this performance *average*.

---

* The true cost of oil may be much greater than this, according to R. Stobaugh and D. Yergin. In their book *Energy Future* (Random House, 1979) they distinguish between the apparent cost of oil and the true cost. They conclude that when proper account is taken of political and economic consequences of consumption of crude oil, the true cost is "... about three times the average price paid by domestic refiners ...".

I estimate that such a device, when used in normal manner throughout the winter, will reduce the heat-loss here—for the winter as a whole—by 34%. This figure is arrived at by assuming:

The shutter or shade is used only during the six coldest months. Thus the winter degree-day figure must be multiplied by, say, 95%.

During those months the device is in use only at night, i.e., from 5:00 p.m. until 7:00 a.m.—a 14 hr period. This period is 58% of the 24-hr day.

At night the indoor vs outdoor temperature difference is about 125% of the difference for the 24-hr day as a whole.

The product of the last three percentages is (95%)(58%)(125%) = 69%.

Multiplying by 50%, which is the effectiveness of the postulated average shutter or shade, one obtains (50%)(69%) = 34%.

Of course, different kinds of shutters and shades may differ widely in thermal performance. Three levels of performance may be defined, as indicated in the table at the top of page 27. Combining the tabulated figures with the 81¢ value of heat loss through 1 ft² of such window in a 5500-degree day location, one obtains 16¢, 27¢, and 40¢ as the cash value of heat saved (at this location) by 1 ft² of a shutter having poor, average, or good thermal characteristics.

Applying the same method to shutters and shades used in a great variety of locations, one obtains the results presented in the succeeding table at the bottom of page 27.

For a 10-ft² window the saving is, obviously, about 10 times as great. For example, the saving provided by an average-performance shutter on a 10-ft² window in Boston is about $2.70 per winter.

For a single-glazed window, the saving is about 2 to 3 times that for a double-glazed window. It may be more if the window leaks much air and the shutter considerably reduces the leakage. But it may be much less if the shutter fails to reduce the leakage.

If auxiliary heat is supplied by electricity the savings are greater, say 1.5 to 3 times as great, depending on the local price of electricity.

| Level of performance (on double-glazed window) | Reduction in heat-loss for winter as a whole when the shutter or shade is applied to the double-glazed window at night only and during 6 months only |
|---|---|
| Poor (When in use, it cuts loss by 30%) | 20% |
| Average (When in use, it cuts loss by 50%) | 34% |
| Good (When in use, it cuts loss by 70%) | 48% |

If the shutter or shade is used in summer also, and reduces the amount of air conditioning used, the annual saving is even greater.

Figures 1–4 present graphs showing the energy savings and dollar savings provided (under certain carefully defined circumstances) by a variety of shutters and shades in various temperature zones.

## MAXIMUM ACCEPTABLE CASH OUTLAY FOR SHUTTERS

To make a reliable estimate of how much a home owner should be willing to pay for thermal shutters or shades is impossible. The estimate would involve many kinds of unknowns, including:

Unknowns concerning the shutters or shades:

How well will they be installed?

How well will they be closed each night? Tightly enough to reduce inleak of cold air?

How many years will they serve? Will repairs be needed? Are they guaranteed?

Unknowns concerning auxiliary heat:

How fast will the price of oil or electricity rise?

Will there be shortages? Rationing?

How much heat will be supplied by the sun, or by a wood stove?

Other unknowns:

What will be the general rate of inflation? How will interest rates change? What Federal and state government subsidies or tax benefits will be available? What tax bracket is the homeowner in? Will better or cheaper shutters be available next year?

The calculation would be difficult and several of the underlying assumptions would probably turn out to be wrong.

I prefer to make an outright guess. My guess is that the cost of oil, electricity, etc., will continue to rise at the rate of 10 or 15% per year and a homeowner should regard the price of a durable shutter

| Location (and degree-day figure) | Cash value* of heat saved per winter, per square foot of well-sealed, double-glazed window, by a shutter or shade the thermal performance of which is: | | |
|---|---|---|---|
| | poor | average | good |
| Los Angeles (2000) | $0.06 | $0.10 | $0.14 |
| Washington, D. C. (4000) | $0.12 | $0.20 | $0.28 |
| Boston (5500) | $0.16 | $0.27 | $0.38 |
| Pittsburgh (6000) | $0.18 | $0.30 | $0.42 |
| Burlington (8000) | $0.24 | $0.40 | $0.56 |
| Duluth (10,000) | $0.29 | $0.50 | $0.71 |

* Assuming that the heat lost is supplied by a 67% efficient furnace burning oil that costs $1/gal.

or shade as being acceptable if it costs no more than ten times the saving during the first year of use. That is, I opt for a ten-times-first-year-saving criterion.

In doing this, I assume that, in fact, the savings will increase each year (as oil and electricity are priced higher and higher) and the shutters or shades will actually pay for themselves in about five years.

Figure 5 presents graphs of maximum acceptable cost of shutters and shades, the criterion being ten-times-first-year-saving.

## SOME GRAPHS OF HEAT SAVING

The following graphs present detailed information on the savings resulting from installation of shutters and shades on well-sealed single-glazed and double-glazed windows.

**Example of use of graphs**  The owner of a house in a 6000-degree-day location contemplates installing an R-5 shutter on a double-glazed airtight window. He has several questions—and their answers may be obtained directly from the following graphs.

1. What percent of the heat lost through one square foot of the window on a cold winter night will be saved by the shutter?
See Fig. 1, lower curve. One finds that an R-5 shutter saves 74%.

2. What percent of the heat lost during the winter as a whole (throughout 24 hours, each day) will be saved by an R-5 shutter that is used during the six coldest months only and during the period 5:00 p.m. until 7:00 a.m. only?
See Fig. 2, lower curve. One finds that an R-5 shutter saves 51%.

3. How many Btus are saved, per square foot of window during the winter as a whole, by the shutter used as indicated above?
See Fig. 3, graph pertinent to 6000-degree-day location, lower curve. One finds that an R-5 shutter saves about 44,000 Btu/ft².

4. How much money (relative to $1/gal. oil) is saved, per square foot of window during the winter

as a whole, by the shutter used as indicated above? Assume the efficiency of the oil burner is 67%.
See Fig. 4, graph pertinent to 6000-degree-day location, lower curve. One finds that an R-5 shutter saves about $0.50 per sq. ft. of window.

5. How much is the homeowner willing to pay for the shutter, if he accepts the ten-times-first-year-saving criterion and if his back-up heat is supplied by a 67% efficient furnace burning $1/gal oil?
See Fig. 5, graph pertinent to 6000-degree-day location, lower curve. One finds that the maximum acceptable cost of shutter is about $5/ft².

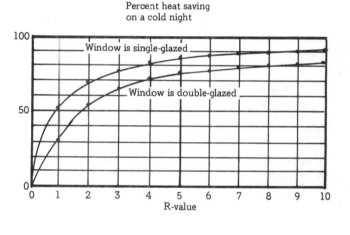

Percent heat saving on a cold night

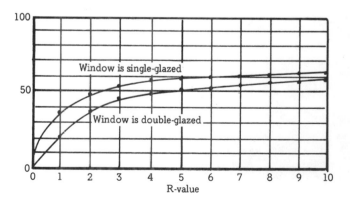

Percent heat saving during winter as a whole

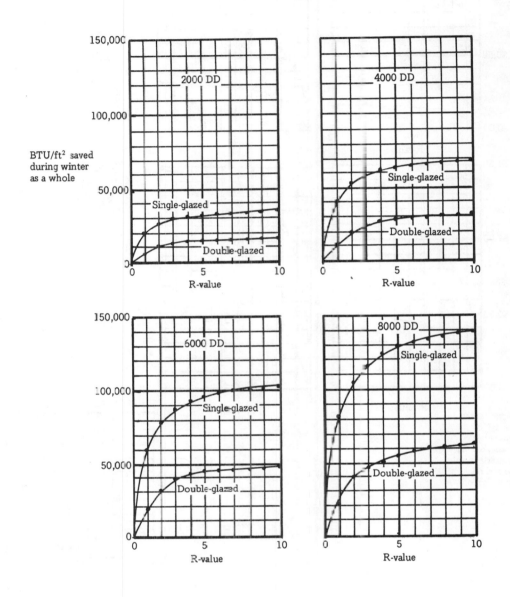

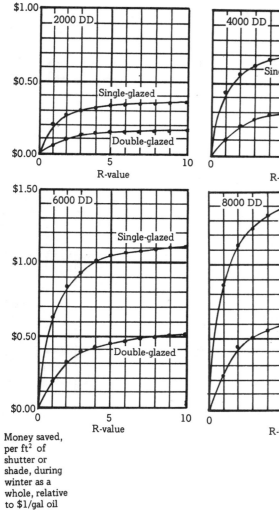

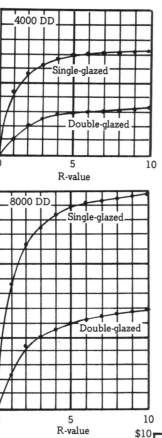

Money saved,
per ft² of
shutter or
shade, during
winter as a
whole, relative
to $1/gal oil

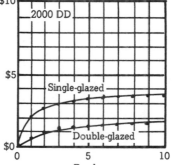

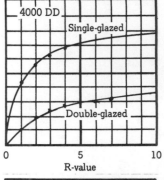

Max. acceptable
cost/ft² of
shutter or shade

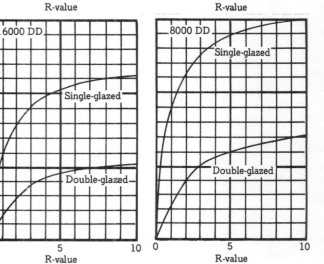

30

**Electrical back-up vs. oil back-up**  Suppose the auxiliary, or back-up, heat is provided by a 95% efficient electrical heating system instead of by a 67% efficient oil furnace burning $1/gal oil. Relative to such electrical back-up heating, how much money do shutters and shades save?

The answer is:

If the electrical energy costs 3.5¢ per kWh, the saving is the same.

If the electrical energy costs 7.0¢ per kWh, the saving is twice as great.

The relationship is so simple that no additional graphs are needed.

What about maximum acceptable cost/ft² when the back-up heat is electrical? Here too the answer is simple. If electricity costs 3.5 per kWh, the maximum acceptable cost is the same as indicated for oil furnace back-up. If electricity costs 7.0¢ per kWh, the maximum acceptable cost is twice as great. For any other cost of electrical power the appropriate answer may be obtained by computing the pertinent simple proportion.

Warning: The accompanying graphs should be used with caution. There are many kinds of circumstances that can make them inapplicable. For example, the shutter or shade may be poorly fitted and its effective R-value may be less than nominal R-value. The window may leak much air, in which case the shutter or shade may be far more helpful or far less helpful than expected, depending on whether it is well sealed or poorly sealed. The outdoor wind speed may be very different from 12 mph. The homeowner may be careless and may fail to keep the shutter or shade closed on many cold nights. Other pertinent factors, called benefits and penalties, are listed in a following section.

## INCREMENTAL COST VS. INCREMENTAL SAVING

When a homeowner is considering buying thermal shutters and finds that various designs having a wide range of R-values are available, and when he discovers that high-R-value shutters usually cost more than low-R-value shutters, he may ask himself: "How high an R-value should I choose? What is the criterion of the most cost-effective shutter? Should I try to maximize the ratio of dollar savings to dollar cost of shutter?"

The fact is, he should try to maximize—not a ratio—but a difference, specifically the difference between (a) dollar savings (over the lifetime of the shutter) resulting from installation and use of the shutter, and (b) the dollar cost of the shutter, including purchase, installation, and maintenance. (Here I neglect added complications relating to interest rates, tax credits, etc.) It is helpful to focus attention on increments: the incremental savings resulting from slightly increasing the R-value and the incremental cost involved in such increase. If the homeowner considers successively larger R-values, and finally arrives at a value such that the two increments (in saving and in cost) are equal, this R-value is the optimum one.

Obviously, in practice one cannot make an accurate and reliable calculation of the optimum R-value. There are too many uncertainties involved, including uncertainties as to how the shutter will perform, how durable it is, how the cost of back-up heat will escalate, etc. Also non-monetary factors are important: appearance, ease of operating the shutter, etc.

## LIST OF MISCELLANEOUS PENALTIES AND BENEFITS ASSOCIATED WITH SHUTTERS

A homeowner considering investing in shutters and shades should bear in mind these miscellaneous penalties and benefits.

### Penalties

The shutters or shades may require maintenance. Eventually they will be discarded. There may be, today, no reliable way of estimating their useful life.

They may cause moisture problems (condensation problems).

They may be esthetically displeasing.

They may present hazards, particularly to children.

The labor involved in opening and closing a shutter each day may be considerable. If the time expended is ¼ minute per day, 160 days/year, this amounts to ⅔ hour per year. If this time is valued at

$6/hour, the labor cost per year, per shutter, is $4. Typically, this slightly exceeds the value of the heat saved by the shutter per winter, if the window is double glazed and has an area of about 10 ft² and if the auxiliary heat is supplied by an oil furnace. If electrical heating is used, the labor cost may be less than the cost of the electrical energy saved.

### Benefits

The annual savings will be twice as great when and if the price of oil (or electrical energy) doubles.

If, later, prices of oil and electrical energy rise so high that installation of shutters and shades becomes imperative, the homeowner would be much better off to have installed the devices today —rather than waiting and finding that the cost of shutters and shades has doubled.

The shutters and shades provide additional benefits, relating to reduction of cold drafts, added security in the event that conventional heat sources are curtailed or cut off, exercise gained in operating the shutters and shades each day, possible use of the devices to help keep rooms cool in summer.

Tax benefits, based on the cost of the shutters or shades, may be provided by the Federal government and some state governments.

With respect to single-glazed windows, the savings provided by thermal shutters and shades are about twice as great as for double-glazed windows.

### DOES THE AVAILABILITY OF FREE HEAT AFFECT THE ECONOMICS?

Yes: if much free heat is available, the cost-effectiveness of shutters and shades is reduced. This is true whether the free heat is supplied by a solar heating system (active or passive) or by other sources.

**Free heat from the sun**  In a house that has an active solar heating system, the larger the collector and storage system, the less the saving from installing shutters and shades. In the extreme case of a house having a solar heating system that provides, even in midwinter, more heat than is needed, no saving at all results from installing shutters or shades. All losses of heat through the windows are made up—free—by solar heat.

Almost every house has south windows and gets at least 10 or 15% of its winter heat need from solar radiation. In such a house, heat-losses via windows in early fall and late spring are made up, free, by solar radiation. However, in the colder months large amounts of heat are provided by the auxiliary heating system and at such times shutters and shades save much heat and money.

In summary, the cost-effectiveness of shutters and shades cannot be computed exactly unless the capabilities of the explicit or implicit solar heating system of the house are known. The greater the capabilities, the less the benefit from installing shutters and shades.

Further discussion of this topic may be found in the article by Dean and Rosenfeld (D-115). They stress that a prerequisite to evaluating the cost-effectiveness of added insulation is a study of the properties of the house as a whole. See, also a report by McGrew et al. (M-94) showing the net gains or losses by south, west, north, and east windows each month throughout the year.

**Free heat from other sources**  The larger the number of people in a given building, the smaller the benefit from shutters and shades. If 100 people were crowded into a typical small house, they themselves would produce enough heat (of the order of 300 Btu per person per hour, i.e., of the order of 700,000 Btu per 24-hr day by the group as a whole) so that no oil or electrical energy would be needed. Here, installing shutters and shades will produce no benefit at all.

Similar arguments apply when use is made of electric light bulbs, electric toasters, washing machines, etc. The more heat these devices produce, the less the benefit from shutters and shades.

If there is a wood-burning stove in the house and if the wood can be obtained free, the cost-effectiveness of shutters and shades may be low.

**Relevance of number of windows equipped**  The saving resulting from installing shutters on one double-glazed south window of a given house exceeds that from similarly equipping a second such window, or a fifth or tenth. As the homeowner equips successive windows, the cost effectiveness of successive installations decreases. As successive windows are provided with shutters, the successive amounts of saved heat are alike but the successive amounts of saving of oil (or gas or electricity) are progressively smaller. Why? Because the greater the number of windows equipped, the greater the fraction of residual heat-loss that can be made up by free heat. (I first learned about this matter from S. C. Baer and obtained additional helpful information from D. A. Boyd.)

Corollary: The better you insulate your house, the less benefit you will derive from shutters and shades.

# CHAPTER 4

# CONSIDERATIONS GOVERNING DESIGN OF SHUTTERS AND SHADES

- Brief List of Criteria of Success
- How Great a Reduction in Heat Loss Should One Seek?
- Use of Trapped Air
- Seals
- Optical Properties
- Convenience
- Durability
- Warping
- Moisture
- Danger of Overheating the Shutter or Glazing on Hot Summer Day
- Sudden-Temperature-Change Danger to Thermopane
- Fire Hazard
- Other Hazards
- Chimney Effect
- Wind
- Reducing the Wind Speed
- Principles Relating to Manufacture, Shipping, Installing, and Adjusting
- Which Location Is Best: Outdoors, Between Glazing Sheets, or Indoors?
- Why Not Design the Devices so That They Can Be Kept on the Windows During the Daytime Also?
- Why R-Values of Shutters and Shades Are Seldom Mentioned in This Book
- Proper Measure of Thermal Performance of a Shutter or Shade

Here I discuss the main design considerations, including criteria of success and principal dangers. Most of the considerations are fairly obvious.

*Note concerning weather stripping and caulking:* Weather stripping and caulking are outside the scope of this book. Detailed advice may be found in many books on insulation and home maintenance. Ordinarily, a homeowner should do any necessary weather stripping and caulking before making a selection of thermal shutters and shades.

## BRIEF LIST OF CRITERIA OF SUCCESS

First and foremost, the device must significantly reduce heat-loss.

Also it must be:

cheap

attractive

easy to operate

durable

easy to install

easy to inspect, clean, repair

unlikely to cause moisture problems

safe with respect to fire, chemical toxicity, mechanical dangers, etc.

Certain other criteria sometimes apply. They relate to (1) acting as reflectors to direct additional radiation into the rooms in winter, (2) serving in summer to exclude solar radiation. See detailed list of special benefits that shutters and shades may provide, p. 2.

One might expect the R-value of a shutter or shade to be a decisively important measure of its success. I believe that this is incorrect, for reasons spelled out on pages 44 and 45.

## HOW GREAT A REDUCTION IN HEAT-LOSS SHOULD ONE SEEK?

Should the designer of a thermal shutter or shade try to cut winter-night heat-loss through the window in question by 20%, 50%, or 90%?

There is no generally applicable answer.

If he can achieve a big reduction in heat-loss, and if the device he designs is durable, cheap, and easy to install, he should indeed seek a big reduction. Otherwise he should be content with a modest reduction.

It seems obvious that if the house in question is in a very cold climate and the windows are single glazed, the designer should attempt to cut the heat-loss by, say, 60 to 90%. But if the house is in a moderate climate and the windows are double glazed, the designer's goal should be only 20 to 60%.

But no hard-and-fast rules can be given. Many special circumstances can affect one's choice of goal. For example:

Big reduction in heat-loss should be attempted if

the house is in a region having very cold winters

the windows are single glazed

there is much in-leak of cold outdoor air (or out-leak of room air)

the costs of oil, gas, electricity, etc., are very high

the windows are large.

A small reduction is likely to be more cost-effective if

the house is in a region having moderate climate

the windows are double glazed

the windows are fairly air tight

low-cost fuel is available

the windows are small (The cost-per-square-foot of a shutter or shade for a small window is likely to be greater than that for a large window.)

the house is well insulated

much free heat is available, e.g., from sun, wood stoves, etc.

**Fallacious Rule-of-Thumb**  Consider this rule-of-thumb: make the window match the wall. That is, provide a shutter or shade that will raise the R-value of the window to equal that of the wall.

This rule is fallacious on several counts. (1) The R-value of the wall is not relevant; the amount of heat that a shutter or shade will save is a separate subject and in deciding what device is most cost-effective the designer should ignore the R-value of the wall. (2) The effective R-value of a shutter or shade is a very tricky and obscure subject unless the window (or the shutter or shade) is essentially perfectly sealed—which is very seldom the case.

## USE OF TRAPPED AIR

Often, the cheapest way of providing much insulation is to make use of trapped (immobilized) air. To trap the air, one may employ:

1. micro-compartmentalization, e.g.: isocyanurate foam, urethane foam, Styrofoam, beadboard

2. macro-compartmentalization, e.g., bubble-type plastic sheet, honeycomb between plastic sheets, honeycomb between glass sheets, thin film of air, large pockets of air

3. myriad fine fibers, e.g., fiberglass, cloth, quilt

4. myriad small beads, e.g., polystyrene-foam beads, hollow ceramic spheres.

Regarding pockets of air: A vertical layer, or pocket, no more than ½-in. thick serves almost as well as a rigid foam sheet of the same thickness. But if the vertical layer is much thicker, convection occurs readily within it, with the result that the R-value may be no greater than that of a thinner layer.

One can use two or more pockets in series—separated by, say, a thin sheet of plastic. The R-value of the sheet proper is very small and of no importance; what is important is that the sheet helps immobilize the air near it.

Better yet: use a thin sheet of shiny aluminum instead of a sheet of plastic. This greatly increases the R-value as explained in Appendix 3.

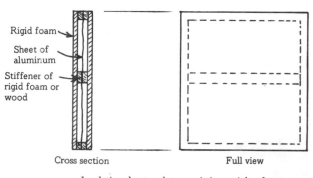

Rigid foam  
Sheet of aluminum  
Stiffener of rigid foam or wood

Cross section          Full view

Insulating shutter plate consisting mainly of two thin layers, or pockets, or air

## SEALS

**The four degrees of tightness of seal**  In designing a shutter or shade, one gives much attention to fit or seal, i.e., how snugly the device will cover the window in question. Almost always there is some air between device and window, and the designer intends that this air be immobilized so that it will help reduce heat-loss.

In designing a seal, one may have in mind any of four different degrees of tightness of seal:

*Degree 1:* (most stringent): The seal is so tight that, even if the shutter remains in place for weeks at a time, no appreciable amount of humid room air will find its way to the (cold) glazing. Thus there can be no build-up of moisture there. No gradual rotting of sills will occur.

*Degree 2:* The seal is just tight enough so that during the course of one cold night relatively little moisture will accumulate on the glazing—so little that it will all evaporate and disappear during the subsequent sunny day with the shutter removed.

*Degree 3:* The seal is just tight enough to reduce by a factor of 2 the amount of air leaking through the window assembly as a whole. If the window assembly itself is already at least moderately tight, the seal of the shutter will have to be about equally tight in order to qualify.

*Degree 4:* (least stringent): The seal is just tight enough so that the amount of room air circulating between shutter and glazing will not be great enough to cause a convective heat-loss greater than the (conductive) loss through the shutter.

**Avoiding waste through over-design of seals**
Clearly, it is a waste of money to provide a superb seal if the window in question already is very tight and room humidity is so low that there is no threat of accumulation of moisture. Why, in buying a shutter or shade, pay $50 extra just to have a tight seal if no tight seal is needed?

Sometimes a tight seal is essential. It would be a waste of money to equip a leaky window with a shutter made of high-R material and provide no tight seal.

A general rule is that (a) high-R material and (b) tight seal do not necessarily go together. Sometimes just high-R material is essential; sometimes just a tight seal; sometimes both.

Companies selling thermal shutters and shades should stock a variety of designs to suit the variety of tightness-of-seal requirements.

Also, I think that such companies, when claiming their seals to be tight, should specify which degree of tightness they mean.

**Possible misuse of seals**    In some houses, seals of shutters or shades are misused: the homeowner uses the seals primarily to reduce in-leak of cold outdoor air, whereas the cost-effective action would have been to seal the window itself. To weatherstrip a window may take 30 minutes. To caulk it may take only 10 minutes. But to provide near-perfect seals for a shutter or shade may be a major undertaking.

**Cosmetic seals**    The seal provided on a commercial shutter or shade may, at first glance, appear very impressive. It may look carefully planned and robust, and may even appear fairly snug. Yet, in fact, it may be largely for looks—cosmetic. The rate of airflow past the seal may be much greater than one might expect and may make the seal almost useless. Accumulation of moisture may still occur. In-leak of cold outdoor air may be slowed only slightly. However, circulation of room air into the space between shutter or shade and the glazing may be stopped.

**Does a strip of felt or pile provide a good seal?**
My guess is that such a strip is adequate for preventing gross circulation of room air into the space between glazing and shutter or shade but is only marginally adequate for preventing accumulation of moisture or preventing in-leak of cold outdoor air.

**Ascertaining tightness of seal**    If a flexible shade is truly well-sealed, sudden application of pressure to one face of it, as by pushing with a large pillow or large breadboard, will fail to displace it far and fast; a strong back-pressure will be felt, and large deflection of the shade will occur only very slowly, i.e., over one minute, say.

If a rigid shutter is well sealed, it will strongly resist sudden opening. There will be a transient "vacuum" that prevents sudden opening.

Of course, if one can detect (with the hand, or with the flame of a candle, or with a plume of smoke, or some dangling threads) jets of air entering each time there is a strong blast of wind outdoors, one knows immediately that the seal is poor. Likewise if one finds, each morning, an accumulation of moisture or ice on the glazing, one knows that the seal is poor.

**Face seal, overlap seal, edge seal**    A *face seal* is made by pressing an insulating plate against a glass pane. Air cannot circulate between the plate and the glass because there is no intervening space.

An *overlap seal* may be made by pressing the plate against the frame of the window sash. There is a ½-in. or 1-in. space between the plate and the glass, but because there is a seal at the top, bottom, and sides, no air can circulate into this space—

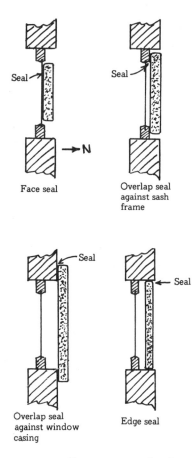

Face seal

Overlap seal against sash frame

Overlap seal against window casing

Edge seal

which may actually increase the heat-saving. Alternatively the overlap seal may be with respect to the fixed frame of the window; the region of trapped air may be several inches thick, i.e., thick enough to allow some (undesirable) circulation of the air within this space.

An *edge seal* is achieved by fitting the plate tightly within the (indoor) window recess, so that each extreme edge bears against the fixed frame of the window.

**What are the tolerances on seals?** Detailed answers are given in Appendix 2.

Some important answers—applicable to a very tight, double-glazed window of average size and an indoor high-R insulating plate—are as follows:

Suppose the plate is to be installed close to the glazing, so as to form a face seal with it.

The plate may be as far as ¼ in. from the glazing without appreciable reduction in effectiveness.

Suppose that the plate is to be installed close to the sash frame, so as to form an overlap seal with it. The plate may be as far as ⅛ in. from the sash without appreciable reduction in effectiveness.

Suppose that the plate is to overlap the fixed frame of the window and is to be many inches from the glazing. Here it may well be that a ⅛ in. gap at the overlap significantly reduces the effectiveness. Effort should be made to keep the gap less than ⅛ in.

In summary, in a considerable variety of circumstances the tolerances are fairly generous. Rumors that seals must always be very tight are not true. (But if a large amount of air leaks through the window, use of a very tightly sealed shutter or shade may be highly desirable. The tightness of the shutter or shade then serves in lieu of tightness of the window itself.)

For outdoor shutters applied to buildings in windy locations, the tolerances are, presumably, much smaller. Very snug fit is essential.

## OPTICAL PROPERTIES

Designers of thermal shutters and shades usually gravitate to materials that are opaque because (1) most objects that have high R-value and are rugged and cheap are opaque (examples: rigid foam, wood, fiberglass), and (2) such materials insure nighttime privacy.

Translucent and transparent devices have, however, this great merit: they can be left in place, covering the windows night and day, week after week. They require no attention.

The face of an outdoor fold-down shutter that is to be used on sunny winter days to reflect radiation toward the window should be aluminized.

Designers who opt for a thin shade find much merit in aluminum films or aluminum coatings because of their high reflectance of far-IR radiation.

## CONVENIENCE

Convenience in installing or removing the shutters, curtains, etc., is important. The occupants wish to be able to close or open the devices in a few seconds, with little physical or mental effort.

Allowing the occupants a choice as to where to store the shutters etc. during sunny days may be undesirable. Some people dislike making decisions, even when the decisions are easy and unimportant. They want to be told: "Put the shutter exactly here." If a shutter can be installed with either face toward the outdoors, they want to be told: "This face should be toward outdoors."

Obviously, if the devices can be left alone for many days, this makes for especially great convenience. A device that is transparent and can be left in place night and day, all winter, is about the ultimate in convenience.

## DURABILITY

Shutters, shades, etc., should be designed for 10 or 20 years' use, if the initial capital outlay is large. If the outlay is trivial, a life as short as one year may be tolerable.

In planning for durability, the designer should take into account:

normal use, including periodic cleaning (washing, e.g.)

prolonged exposure to high temperature, especially in summer

prolonged exposure to sunlight

rough handling by children

prolonged exposure to moisture resulting from condensation

effects of rain, snow, wind, etc.—if the device is mounted outdoors.

Roll-up shades present special problems, such as scuffing of the edges, skew roll-up, failure of the spring return or of the pawl.

If a removable shutter is held closed by buttons or latches, there is the possibility that, when the occupant removes the shutter, he will pull it away from the window before all of the buttons or latches have been released. In such case excessive stress may be exerted on the overlooked button or latch, and this device may be bent or broken, or the pertinent portion of the shutter may be bent. Such danger can be reduced by (1) use of very few buttons or latches, (2) use of very shallow engagement so that the device releases the shutter as soon as the shutter is moved even slightly, (3) making the button or its axle of compliant material.

## WARPING

**Cause of warping**   A shutter may warp because the two faces are of different materials and, as time passes, one face shrinks (or expands) more than the other. Asymmetry of the irradiation and heating in winter or summer and asymmetry of moisture accumulation may contribute to warping. Mechanical stresses associated with installing the shutter, or clamping it in place, or parking it in sloping orientation, may produce warping.

**Harmfulness of warping**   Certain geometries of warping (of a typical shutter that is secured by tabs, ears, or the like) tend to be helpful, but most geometries tend to be harmful, i.e., they produce cracks that permit unwanted circulation of air.

If the warping of an indoor shutter opens a gap at the bottom, the colder air between shutter and window can escape via this gap. The opposite applies to an outdoor shutter; a gap at the top is harmful, because the air between shutter and window is warmer than outdoor air and accordingly is less dense, and accordingly will escape via the gap.

The following sketches show harmful geometries of warping of indoor shutters.

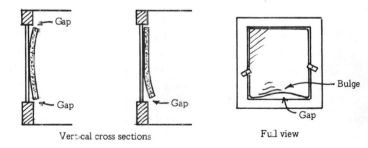

Vertical cross sections                    Full view

**Corrective procedure** Harmful warping could conceivably be avoided by imparting to the plate, in advance, a tendency to warp in the opposite manner, i.e., in a manner that will be opposed by the buttons, tabs, etc., provided.

Consider, for example, a shutter to be held by two buttons midway up the sides. Then an appropriate procedure is to impart to the plate a tendency to become cylindrically curved with the convex side toward room and with the high-points of the dome along the line from button to button. When such shutter is secured against the window by this pair of buttons, good edge seals all-around will result.

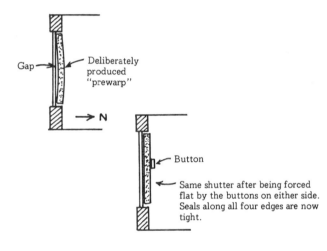

Vertical cross sections

There are many ways of creating, at the factory, a small extent of helpful warping;

Build the shutter in a jig that has the desired warp. On being removed from the jig, the shutter will retain this warp.

Build flat shutter, then subject it to heat while exerting forces tending to produce the desired warp.

Produce the desired warp by applying, to one face of the plate, coatings of special chemicals that tend to make the material of the plate shrink (or expand). For example, if one sprays a tough plastic coating on almost any kind of flat sheet, the sheet will later be found to have acquired a slight curvature—due to the differential expansion or contraction of the coating.

If a shutter warps harmfully, one should consider reversing it, so that the warp will be helpful.

**Brute force tactics** One brute force tactic for avoiding harmful warping is to incorporate in the shutter a framework of stiff wooden or metallic strips. It may suffice to make just the perimeter (edges) very stiff.

Another brute force method is to make the shutter thicker. Thicker objects warp less than thin ones.

## MOISTURE

When thermal shutters or shades are installed, build-up of moisture may occur and may be troublesome. Warm air may reach a very cold surface and moisture may condense there, especially on cold nights. Water droplets may be observed, or tiny puddles, or an accumulation of frost. During a sunny day such water may evaporate and disappear—but this process may be inhibited by the shutter.

If the shutters or shades are mounted outside the windows, i.e., on the outdoor side, the threat of moisture build-up is small. In winter the absolute humidity of outdoor air is low and the window glass is usually warmer than the outdoor air. Thus moisture build-up is not likely to occur, and if it does occur the moisture is likely to evaporate before noon on the next sunny day.

If the shutters or shades are mounted between the glazing sheets, the threat of moisture build-up can, under some circumstances, be severe. This matter is taken up in Chapter 11.

If a certain type of shutter is pervious to water, and this contributes to a moisture problem, painting the shutter with an impervious paint might be worthwhile. For example, Insul-Aid paint is said to have a water permeability of 0.6—far less than that of most other paints (but not as low as that of a polyethylene vapor barrier).

If a designer fears that accumulated moisture may rot the wooden sills, he may employ rot-resistant wood (cyprus, redwood, or cedar), or specially treated wood, e.g., wood that has been treated with pentaphenol or cuprinol. (Reference: *Solar Age, June 1979, p. 7.*)

## DANGER OF OVERHEATING THE SHUTTER OR GLAZING ON HOT SUMMER DAY

If a shutter is in an enclosed space and is black, it may become very hot on a hot sunny day in summer—so hot that it suffers permanent damage from excessive thermal expansion or from partial melting. An adjacent sheet of enclosed glass may become very hot and, if it lacks sufficient expansion space, may crack. (I am indebted to Bruce Anderson for pointing out these dangers.)

The dangers may be avoided by providing a large roof overhang that will shade the window in summer or by using a shutter that has a white or aluminized surface. The danger is less for a vertical window than a window that is sloping downward toward the south.

## SUDDEN TEMPERATURE CHANGE DANGER TO THERMOPANE

Suppose that, on a very cold night, a highly effective indoor thermal shutter is installed in a window consisting of Thermopane, Twindow, or other rigid assembly of two spaced sheets of glass. Suppose that, a few hours later, the shutter is removed. Just before the shutter is removed, both sheets of glass are very cold. Soon after the shutter is removed the inner sheet becomes nearly as warm as the room, and accordingly the two sheets are at very different temperatures. There is a chance that the differential thermal expansion may cause breakage. (I am indebted to J. W. Restle of Ark-Tic-Seal systems, Inc. for pointing this out.)

If an outdoor shutter is used, the danger may be even greater, because when the shutter is removed the cold outdoor wind may cool the outer glass sheet especially rapidly and the differential thermal expansion may be especially great.

## FIRE HAZARD

If feasible, the materials used in constructing the shutters, shades, etc., should be non-flammable.

In any event they should not be highly flammable. Preferably they should be no more flammable than other objects of comparable size and accessibility in or near the rooms in question.

The materials should be such that, when involved in a fire, they will not emit appreciable quantities of highly toxic chemicals.

Materials may be painted with fire-retardant paint. In painting Styrofoam or other materials that may be dissolved by ordinary paints, use a latex-base fire-retardant paint, such as:

"Flame sheet 76-Line" made by M. A. Bruder & Sons, 600 Reed Rd., Broomall, PA 19008, or

"Retardo" made by Benjamin Moore & Co., 51 Chestnut Ridge Rd., Montvale, NJ 07645, also 134 Lister Ave., Newark, NJ 07105.

## OTHER HAZARDS

The designer must keep in mind:

Mechanical hazards are:

Sharp prongs, sharp edges, etc., which could damage the hands or eyes.

Levers or arms that might swing and hurt someone.

Falling heavy objects.

Strings, loops, brackets which could choke a child.

Chemical hazards, including chemicals that could poison people who touch the materials in question and then lick their fingers. Children might lick the materials directly.

Electrical hazards, if electric wires are employed.

Dangers associated with living organisms, e.g., moulds, fungi, bacteria that may grow in confined regions where moisture has accumulated.

Tendency of bees, wasps, etc., to build nests in small outdoor spaces that are protected from the rain.

It is to be expected that, in most houses, children will sometimes play with the shutters or shades. For example, children may climb onto the shutters, or hang on them, or swing from them.

For a detailed discussion of many of the hazards involved, see Ref. S-65.

## CHIMNEY EFFECT

Can poorly arranged curtains actually increase the heat-loss? They can, according to an article on p. 13 of the June, 1977, *Bulletin of the New Mexico Solar Energy Association*. It is claimed that the curtain, if spaced some distance from the window, can form a chimney-like region in which the vertical flow of air by gravity convection is encouraged, with consequent increase in rate of heat-loss. However, no convincing evidence was presented and inquiry failed to elicit convincing evidence. I remain skeptical.

## WIND

A shutter or shade situated outdoors must be able to withstand wind—not just typical wind but the highest wind to be encountered in a decade.

## REDUCING THE WIND SPEED

Inasmuch as reduction in wind speed reduces heat-loss (see Chapter 2), the designer may find it worthwhile to employ a scheme for reducing the wind speed, i.e., slowing the air that moves close past the outer faces of the windows.

Just how strongly the wind speed affects the rate of heat-loss through windows is made clear in Chapter 2. When the speed is very low (below 3 or 6 mph), reducing the speed further produces large percentage reduction in heat-loss; but in absolute terms the reduction is small. When the speed is high (above 12 mph), the effective thickness of the insulating airfilm just outside the outer glazing is nearly zero; therefore cutting the speed in half produces little reduction in heat-loss (about 15% reduction, typically); but a tenfold reduction in wind speed can be very helpful.

Some ways of reducing the speed of the outdoor air that is very close to windows are:

Place the house in a valley or in deep woods.

Plant a windbreak of trees.

Ascertain the direction of the prevailing wind and arrange to have many windows on the leeward side of the house and few on the windward side.

Recess the windows deeply into the side of the house and, near the edges of the recess, install damping devices (see below).

Install, 2 to 10 ft. upwind of the windows, large, air-pervious devices (with 50% porosity) that will slow the wind (damp it) without producing vigorous whirls of wind. Use, for example, baffles that resemble snow fences, i.e., have alternating obstructions and open passages. A stack of nettings serves well. Or a stack of sheets of expanded metal. Or a loose bundle of wooden strips. Or a bundle of slender and irregular branches of trees.

## PRINCIPLES RELATING TO MANUFACTURE, SHIPPING, INSTALLING AND ADJUSTING

The manufacturer should seek schemes that:

Permit a few general types of devices to be applicable to a wide variety of types of windows.

Permit a few sizes to be adaptable to a wide variety of window widths and window heights.

Permit rough handling without danger of damaging the devices.

Permit compact shipment, preferably without need for crates or cartons.

Permit the size to be adjusted at the house. If, on arriving at the house, the installer finds that the windows are of different size than expected, he will want to alter the size of the shutters or shades immediately. Easy on-site adjustment is a great virtue.

Permit the homeowner himself, with only the simplest of tools, to make whatever adjustment is necessary. Shutters that satisfy this requirement could be sold at supermarkets, with the entire burden of delivery, adjustment, and installation shifted to the buyer. As a consequence, the distributor could set his price very low.

## WHICH LOCATION IS BEST: OUTDOORS, BETWEEN GLAZING SHEETS, OR INDOORS?

The device for reducing nighttime heat-loss can be placed just outside the window, or between the two sheets of glazing, or on the indoor-side of the window.

Which of these locations is best? There is no single answer. In different situations, different locations are optimum.

One might be tempted to conclude that all three locations are equally good. But such conclusion is not quite valid. And, if certain practical matters are taken into account, it is far from true. That is, in some situations an outdoor location may be far superior and in other situations another choice of location may be far superior.

### Outdoor Location

#### Advantages:

A shutter of large area can be used. (Large indoor shutters may be objectionable: they may be too hard to put out of the way.)

The shutter can made to do double or triple duty. Besides insulating the window on winter nights and providng privacy at night, it can be used to reflect additional solar radiation into the window during winter daytime, if the shutter is suitably oriented and its upper face is aluminized; and it can be used in summer to exclude solar radiation, if the shutter is tilted so as to act like an awning or overhanging eaves.

Any moisture that exists in the space between shutter and window tends to go away, inasmuch as the outdoor humidity in winter is very low and the space in question communicates with outdoors.

#### Disadvantages:

The shutter must be able to withstand hurricane winds, driving rain, heavy snow, intense sunlight, nesting bees, squirrels, etc.

The residents have to go outdoors to operate the shutters—unless there are control rods or control ropes that pass through the wall.

The shutter may detract from the outside appearance of the house.

The shutter may interfere with trees, shrubs, etc.

### Location Between the Glazing Sheets

#### Advantages:

The device occupies no space outdoors and no space indoors. The space in question is used for no other purpose.

The device is relatively inaccessible, hence can be thin and delicate.

Little dust can reach it. It may remain clean for years.

Air currents cannot reach it. It will not blow about. It does not have to be strongly held.

#### Disadvantages:

Access to the device, as for servicing it, is difficult.

In some windows there is little space between glazing, and accordingly only a small reduction in heat-loss is possible.

Moisture may accumulate on the device and may remain there a long time.

### Indoor Location

#### Advantages:

The residents can close or open the device without having to go outdoors.

When the device is opened, any moisture that was trapped there will soon evaporate (on a sunny day, at least).

The device does not have to withstand wind, rain, snow, bees, etc.; thus high-quality seals are not needed, and the attachment devices can be weak and informal.

The exterior appearance of the house remains unchanged.

**Disadvantages:**

The appearance of the room is altered.

When the device is closed, moisture may built up between it and the window.

The device must be designed with much regard to hazards such as fire, release of toxic chemicals, mechanical injury, etc. It must not pose a threat to children.

If the shutters are of detachable type, finding places in which to "park" them may be difficult.

## WHY NOT DESIGN THE DEVICES SO THAT THEY CAN BE KEPT ON THE WINDOWS DURING THE DAYTIME ALSO?

This is an excellent question. The writer is convinced that there is much merit in doing exactly this. Transparent or translucent sheets can be used and can be left in place day and night. The merits and demerits of such schemes are discussed in a later chapter.

## WHY R-VALUES OF SHUTTERS AND SHADES ARE SELDOM MENTIONED IN THIS BOOK

It may seem surprising that I seldom state R-values of shutters or shades. I state R-values of materials but seldom state R-values of complete shutters or shades.

Why not? Because:

- R-value is not defined for an actual shutter or shade in use on an actual window. Strictly speaking, R-value applies only to a uniform plate that has no edges (or the edges are perfectly sealed or are somehow perfectly compensated). It is not defined with respect to a plate that is of limited size and has edge gaps that allow passage of air.

- R-value is a single number, yet an actual shutter installed on an actual window reduces many kinds of heat-losses and may reduce each to a different extent. The losses include:

loss by radiation,

loss by conduction,

loss by convection of air within the room,

loss by out-leak of warm room air through:
  cracks around the glass panes,
  cracks around the sashes,
  cracks in the window frame as a whole,

loss associated with in-leak of cold outdoor air through such cracks.

Clearly, a single number cannot characterize the ability of a shutter or shade to reduce heat-loss through a window of unspecified type, especially if nothing is known as to how leaky the window is, the speed and direction of the wind, and the distribution of indoor air currents.

- The performance of a shutter or shade may fall below that predicted by the manufacturer if the device was installed incorrectly, or if it is not closed correctly at night, or if the device degrades physically after a few months or years of use.

- The worth of such device is not proportional to the R-value. Beyond about R-5 the law of diminishing returns sets in strongly.

- The goal of the user of such device is not to achieve high-R but to achieve high overall saving (and ease of operation, attractive appearance, etc.) Some of the devices that are most cost-effective have low R-values—such as 2 to 5.

- To determine the actual R-value of an actual device installed on an actual window is very difficult. For most home owners it is virtually impossible.

- Many of the values stated by manufacturers are based on calculations, and many of the calculations are made with little regard to air-leakage through the window and air leakage around the shutter or shade. Some of the values stated are exaggerated. It is not feasible for me to try to decide which values are exaggerated and which are correct.

- Some manufacturers give R-values that apply not to the shutter or shade itself but to the combination of such device and the window. Users of such devices may be confused by this type of specification.

**Effective R-value**    Would it be helpful to define an effective R-value of a shutter or shade, i.e., a value that would take into account the tightness of fit, the extent to which air leaks through the window, etc.?

In some special circumstances, yes. But in general, no. The trouble is the effective R-value changes widely with circumstances. Consider, for example, a thick Styrofoam shutter. If it is applied poorly (large gaps at the edges) to a perfectly sealed window, the effective R-value of the Styrofoam may be high (8, say). But if it is applied to a very leaky window the effective value may be as low as 2. Conversely, consider a 0.001-in. film of the cheapest plastic. Applied to an airtight window it may contribute only R-½. But applied to a very leaky window and perfectly sealed to it, it may reduce heat-loss by 90% and deserve a much higher effective R-value.

Note concerning R-values of walls:  I have recently learned that some experts on very-high-quality insulation for walls have become unhappy about stating the R-value of a wall. Such values are usually calculated with respect to an ideal portion of the wall; yet poor seals along the edges of the wall, and cracks between butt-jointed insulating plates, may permit air leakage much greater than expected. A wall rated R-15 may perform no better than a wall that truly provides R-10. Again we are reminded that, strictly speaking, R-values apply only to "edgeless" sheets.

## PROPER MEASURE OF THERMAL PERFORMANCE OF A SHUTTER OR SHADE

If R-value is not a proper measure of the thermal performance of a shutter or shade, what is a proper measure?

Answer: the resulting heat-saving, i.e. the saving on a typical night in winter or the saving for the winter as a whole. The former is easier to evaluate. Evaluation may involve calculation or measurement, as explained below.

**Calculation:**    Calculate the effectiveness of each layer: each individual sheet or plate and each air-film and then calculate the effectiveness of the combination.

Then look into the edge gap and crack situation. See what gaps there are at the edges of the shutter or shade, and what cracks there are in or around the window. Find what the typical speeds and directions of winds are, adjacent to the particular house in question (and taking into account nearby windbreaks consisting of houses, trees, shrubs, hills, etc.). Estimate what in-leak or out-leak of air will occur (a) without the shutter or shade and (b) with it.

Estimate the heat saving attributable to the shutter or shade.

Alas the result may be more like a guess than a calculation!

(Suggestion: Eliminate air leakage through the window by caulking, etc. This greatly simplifies the calculation. In addition, it saves heat!)

**Measurement**    Measure the normalized between-glazings temperature head in the window in question. Do so: (a) without the shutter or shade, and (b) with it. The procedure is explained in Appendix 1.

# CHAPTER 5

# MAIN INSULATING MATERIALS

- Thermax
- Polystyrene Foam
- Urethane Foam
- Thermoply
- Fiberglass
- Foamglas
- Wood
- Aluminum Foil
- Other Opaque Flexible Sheets

Here I describe opaque insulating materials, rigid and flexible, used in thermal shutters and shades. I describe just the main materials. For a more complete list, see Appendix 5. Translucent and transparent materials are discussed in the following chapter and in several appendixes.

Some of the materials present hazards, discussed in Reference S-65.

## THERMAX

This material, made by Celotex Corp., a subsidiary of Jim Walter Corp., consists of a plate of closed-cell thermosetting isocyanurate foam (with some added fiberglass, for strength) and, on both faces, 0.002-in. aluminum foil. Various overall thicknesses from ½ in. to 1½ or 2 inches are available. Sheet size: 8 ft × 4 ft. Density: about 2.5 lb/ft$^3$. R-value: about 8 per inch. Retail cost (1979) of a ½-in. sheet: about $0.26 per ft$^2$.

The sheets are surprisingly stiff and strong. They show little tendency to warp. The aluminum faces are attractive. The material may be cut with a sharp knife or—better—with a carpenter's saw, which advances about 4 inches per stroke. Some fiberglass is evident in exposed edges, making it desirable to apply tape to the edges. Masking tape adheres firmly, but other kinds of tape (e.g., duct tape, or various aluminized tapes such as Tape #488 made by Ideal Tape Co.) may be more durable.

The material is flammable and, in burning, emits poisonous gases. The aluminum foil faces somewhat inhibit or retard combustion.

The plate can be painted with almost any kind of paint. Use of a fire-retardant paint may be desirable. Or it can be covered with cloth or paper of pleasing appearance.

Two small pieces may be butt-joined and secured with tape to form one large piece.

Tape

Butt joint

Note: The *High-R Sheathing* sold by Owens-Corning Fiberglas Corp. is virtually identical to thermax.

## POLYSTYRENE FOAM

Expanded polystyrene (polystyrene foam) is made by several companies, whose products are called Styrofoam, Perma-Foam, etc.

**Styrofoam** This material is made by Dow-Corning Co. Thicknesses from 1 in. to several inches are available. Sheet size: 2 ft × 8 ft. Density: 2 lb/ft$^3$. R-value: about 5 per inch. Retail cost (1978 or 1979) of 1½-in. sheet: about $0.36 per ft$^2$.

The sheets are stiff and strong and show little tendency to warp. They may be cut with a carpenter's saw, which advances several inches per stroke. The edges are not abrasive and do not need to be taped.

The material is flammable and, in burning, emits poisonous gases. It is said to include some fire-retardant.

The material can be painted with latex paints. It may dissolve and shrink when various other kinds of paint are applied.

Type SM, bluish in color, is often used as siding. Type RM is used on roofs. Type TG has tongues and grooves.

Two small pieces may be butt-joined and secured with tape to form one large piece.

**Perma-Foam** This is made by Mid-America Industries, Inc. R-value: about 4 per inch.

**Zonolite Styrene Foam** This material (not to be confused with Zonolite vermiculite) is made by Grace Construction Products, a division of W. R. Grace & Co.

## URETHANE FOAM

This material, made by National Gypsum Co. and called Ze-O-Cel, and also made by Upjohn Co., has properties somewhat similar to those of isocyanurate foam and Styrofoam R-value: about 7 per inch. Density: 2 to 4 lb/ft$^3$. The commonest types (#200, red or tan in color), although containing fire-retardant and called self-extinguishing, are flammable and produce poisonous combustion products. Type 25, which is ivory-colored, is recommended by the manufacturer for ". . . applications requiring noncombustible insulation."

## THERMOPLY

This material, made by Simplex Products Group or by Simplex Industries, Inc., consists of a hard, dense, ⅛-in. cardboard sheet with aluminized foil on one face. The other face has been painted white.

The material shows some tendency to warp and is flammable. its R-value is much lower than that of a ½-in. Thermax sheet.

## FIBERGLASS

This material is made by many companies, including Owens-Corning, Johns Manville, and others.

It is non-flammable, or at least only a small fraction of it (the binder?) is flammable. R-value: about 3.0 to 3.5 per inch.

### Owens-Corning brand:

The commonest kind is pink in color and contains about 9% by weight of organic binder. The binder may outgas if heated to temperatures approaching the boiling point of water, and the products of outgassing can cloud adjacent glass plates.

Type SL-20: this contains only 1½% organic binder. There is less tendency to becloud glass covers.

Type SL-10: this contains no binder and has relatively little cohesiveness or strength. The material has almost no tendency to becloud adjacent glass covers.

## FOAMGLAS

This material, made by Pittsburgh Corning Corp., has many very desirable characteristics. It will not burn. It will not corrode. It is entirely impervious to water. It is very rigid and stiff—will not warp. However, its R-value (2½ to 3 per inch— is only about half that of typical organic foams and its density (9 lb/ft³) is more than three times that of common organic foams.

## WOOD

Various kinds of wood have been used in shutters. R-values of wood are fairly low—about 1.3 per inch. Wood is flammable—and expensive. In the course of a few years it may shrink (in direction transverse to the grain) and warp. If left wet for long periods it may rot.

## Summarizing graph

The accompanying graph summarizes much information on the R-value of various common insulating materials. The values are approximate only: slightly different brands, or formulations having slightly different density, have somewhat different R-values. Also the R-value of a material changes somewhat with the temperature of the material, being lower as the temperature increases.

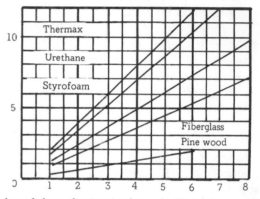

R-values of plates of various insulating, far-IR-absorbing, materials. The values are given for thicknesses up to 2 in. (The R-values of the flanking air films, if any, are not included. Values for very thin plates are not included because such plates may have appreciable transmittance of far-IR and additional energy-transfer mechanisms come into play.)

## ALUMINUM FOIL

The commonest aluminum foil is about 0.001 in. thick and is easily cut with scissors. The foil is easily attached to plates or frames by means of tape. The material is non-flammable unless raised to very high temperatures. The 1979 retail cost was about $0.10 per ft².

When new, the foil has a visual-range reflectance of about 0.9 and a far-IR absorptance of only about 0.05. Aged foil has poorer optical properties: the visual-range reflectance may be only 0.6 to 0.8, and the far-IR emittance may somewhat exceed 0.05.

Designers of thermal shutters or shades often employ aluminum foil that is cemented to a thick rigid plate or is mounted between flexible plastic sheets or quilts. While these procedures protect the aluminum from mechanical damage and may be helpful in other ways, they may largely defeat the ability of the foil to reduce heat-flow-by-far-IR—

radiation. A bare foil—foil flanked by at least ¼ in. or ½ in. of stagnant air on each side is extremely effective in reducing heat-flow.

Why is this? The effectiveness of the airfilm on the room side (north side) of the foil is high because nearly all of the far-IR radiation incident on the north side of the foil is sent back (instantly!) to a location beyond this airfilm; accordingly the foil absorbs practically no energy from the room other than kinetic energy that slowly makes its way through this airfilm by conduction or convection. The effectiveness of the airfilm on the south side of the foil is high because the foil emits practically no radiation; accordingly, practically no energy flows from the foil toward the outdoors other than kinetic energy that slowly makes its way through this airfilm.

If the foil were closely flanked by sheets of material that absorb and emit far-IR radiation (as most things do), there would be no highly effective airfilms. Energy would pass through such airfilm instantly (by radiation process) and would quickly pass through the foil itself by conduction. The high conductive resistances of the stagnant airfilms would, in a sense, be by-passed. A main virtue of the aluminum foil would be defeated. (Of course, the foil may be useful in other ways, such as stopping leakage of air.)

*R-value:* The R-value of an aluminum foil de-

| Situation of foil | Approximate R-value |
|---|---|
| Foil tightly sandwiched between glass, wood, heavy cloth, or other material having high far-IR emittance and absorptance | 0.0 (foil itself) |
| Foil flanked on both faces with a thick region of still air | 2.7 (foil and adjacent air) |
| Foil flanked on one face with air and on the side with glass, wood, heavy cloth, or the like | 1.4 (foil and air on one side) |
| Two foils 0.5 to 3.5 in. apart * | 2.5 (foils and air between them) |

* For more detailed information on two foil sheets and the intervening air, see Appendix 3.

pends greatly on how fresh and shiny the surfaces are, how well the edges are sealed, and what materials lie immediately adjacent to the foil. Some illustrative values are indicated in the accompanying graph. In every case, very shiny surfaces (emittance 0.05) and tight edge seals are assumed.

## OTHER OPAQUE FLEXIBLE SHEETS

Many opaque thermal shades are made of flexible material: cloth or plastic films. Most such shades are put into use at the end of the afternoon and moved out of the way early in the morning. Some such devices may be rolled up, some are slid laterally along horizontal curtain rods, and some are accordion-folded vertically.

Among the most common opaque flexible sheets are:

**Vinyl plastic.** About 0.006 in. thick.

**Vinyl plastic to which scrim cloth has been laminated.**

**Aluminized mylar.** Typically, the aluminum foil is about 0.0005 in. thick and is laminated between mylar films. The combination is strong and tough and (unlike bare aluminum foil) may be rolled up and unrolled repeatedly without losing its smooth, shiny appearance.

However, the reflectance is usually somewhat less than that of bare aluminum. This is because the refractive index difference between aluminum and mylar is less than that between aluminum and air.

Aluminized mylar is made by duPont Co. and by Madico, Inc.

Some other interesting opaque flexible materials are:

**Foylon 7001** Includes 0.0003-in. aluminum foil face bonded to a polyester backing. Overall thickness: 0.004 to 0.008 in. Intended for use as greenhouse thermal blanket and as a drapery liner. Front face is the aluminum face and has clothlike texture. Back face is white. Costs about 45¢/ft² in large quantities.

**Foylon 7137**   Like the above except thicker with black vinyl backing. Costs about 60¢/ft² in large quantities.

**Dura-Shade 4413**   Like the above except thicker yet—about 0.015 in. Includes a vinyl film backing. Costs about $1/ft² in large quantities.

**Astrolon**   Includes vacuum-deposited aluminum film protected by polyethylene films.

Some of the main manufacturers of opaque flexible sheets are:

**duPont Co.**   Aluminized mylar. Also Tyvek (containing polyolefin) and aluminized Tyvek.

**Duracote Corp.**   Foylon, an aluminum-coated fabric.

**King-Seeley Thermos Co.**   Astrolon, a thin aluminum film between polyethylene films.

**Madico, Inc.**   Various highly reflective films, including aluminized mylar. The company makes Reflecto-Shield, Shade-Shield, and Fade-Shield.

# CHAPTER 6

# OUTDOOR TRANSPARENT AND TRANSLUCENT INSULATING DEVICES

- Manufacturers of Glass Sheets
- Manufacturers of Replacement-Type Windows Double-Glazed with Glass
- Manufacturers of Rigid and Semi-Rigid Plastic Sheets
- Manufacturers of Flexible Plastic Sheets
- Manufacturers of Bubble-Plastic and Honeycomb-Plastic Sheets
- Manufacturers of Dual-Wall Translucent Plastic Sheets
- Means of Attaching a Rigid Sheet
- Means of Attaching a Flexible Sheet
- Which Attachment Location Is Best?
- How to Prevent a Large Outdoor Sheet of Flexible Plastic from Flapping and Bellying
- Outdoor Set of Two or More Sheets of Translucent Glazing

An easy way to reduce loss of heat through a large double-glazed window on cold nights is to add another sheet of glazing. That is, make the window triple-glazed. Or add two sheets to make it quadruple-glazed. The sheets may be rigid or flexible, transparent or translucent, permanent or removable. They may be mounted on the outdoor-side of a window, or between glazing sheets, or on the indoor side. In this chapter we deal just with the outdoor mounting.

One may ask: "Is not this book concerned only with shutters and shades, and if so, why is the addition of another sheet of glazing a proper subject to include?"

I include this subject on the slender pretext that the additional sheet may be installed by the homeowner himself, may be installed informally and cheaply, and may serve in lieu of a shutter or shade.

**Extra benefits**   The added sheet of glazing provides extra benefits beyond merely reducing heat-loss at night. It can be left in place during the day to reduce daytime heat-loss. It can be left in place in summer to reduce inflow of heat.

**Drawbacks**   An added outdoor glazing sheet may interfere with opening the window and also with cleaning it. Having a reflectance of about 8%, the sheet slightly reduces the amount of solar radiation entering the room. If the sheet is translucent rather than transparent, it prevents room occupants from obtaining a clear view of the outdoors.

**Special requirements**   Outdoor glazing sheets must be able to withstand:

> prolonged irradiation (visible light, UV, IR) rain, snow, and ice formation
>
> high winds
>
> insects searching for sites for nests

Preferably the sheets should be transparent, not merely translucent.

**How to conceal the fact that an added sheet is translucent, not transparent**   Install white gauzy curtains. Or suspend potted plants or other bric-a-brac in front of the window. Persons inside the room will then hardly notice that they cannot see out clearly.

Leave one small portion of the window truly transparent, so that a room occupant may—if he must—get a clear view of the outdoors.

**How big is the reduction in heat-loss?**   An added transparent or translucent sheet will reduce heat-loss of a double-glazed window only about 10 or 20% if the sheet transmits far-IR readily and if the edge seals are poor.

But if the sheet absorbs the far-IR, and if its edge seals are tight and 1/2 in. or more of air is trapped behind it, the reduction in heat-loss may be 20 to 40%. (If the sheet were to reflect the far-IR, the performance would be much better yet.)

If the window proper is very leaky and the added sheet is not leaky, the added sheet may be highly valuable in reducing in-leak of cold outdoor air.

**How tight must the edge seals be?**   Tightness of seal is discussed in Chapter 4 and Appendix 2.

As regards seals of outdoor glazing sheets, the main fact is that the seals must be very tight indeed, otherwise insects, driving rain, etc., will enter the enclosed space and become a major nuisance. Seals good enough to exclude wind-driven rain may be better than necessary from the viewpoint of thermal insulation.

A small gap near the lower edge of a poorly sealed outdoor glazing sheet does little harm as regards thermal insulation. This is because the region of trapped air communicates mainly with the outdoors, the trapped air is warmer than outdoor air, and warmer air tends to rise, not fall.

## MANUFACTURERS OF GLASS SHEETS

Glass is a near-ideal glazing material. It is transparent, strong, dimensionally stable, non-corroding, air-tight, water-tight, impervious to moisture, non-flammable, and fairly inexpensive. Tempered glass is especially strong.

The drawbacks are: it is heavy, fragile, and has low intrinsic R-value.

Standard thicknesses of small-area sheets for windows:

● single-strength glass—about 3/32 in. (0.085 to 0.100 in.)
● double-strength glass—about 1/8 in. (0.115 to 0.133 in.)
● triple-strength glass—about 3/16 in. (about 0.19 in.)

Some of the main producers of flat glass are:

**PPG Industries Inc.:** *Herculite, Twindow*

**Libbey-Owens Ford Co.:** *Thermopane*

**ASG Industries, Inc.:** *Sunadex, Solatex, Sun-A-Therm*

See also Appendix 6.

## MANUFACTURERS OF REPLACEMENT-TYPE WINDOWS DOUBLE-GLAZED WITH GLASS

Instead of buying additional glass sheets to be attached to a window, a homeowner may buy an entire new window—a replacement-type window that includes two sheets of glass. Some manufacturers of such windows are:

Alcoa Buildings Products, Inc., Grant Bldg., Pittsburgh, PA 15219

Anderson Corp., Bayport, MN 55003

Rolscreen Co., Pella, IA 50219

Season-All Industries, Inc., Indiana, PA 15701

## MANUFACTURERS OF RIGID AND SEMI-RIGID PLASTIC SHEETS

Rigid or semi-rigid plastic sheets are superior to glass sheets in several ways: they are tougher, less brittle, lighter, and easier to cut to size. Also they have an intrinsic R-value about six times that of glass of the same thickness. (Warning: intrinsic R-value of a sheet of glass or plastic is usually very small compared to the effective R-values of the flanking films of stagnant air.)

They are inferior in several ways: they are flammable, may undergo dimensional change, may degrade under prolonged exposure to sunlight or high temperature, and are less transparent (at least after they have been cleaned many times—plastic is easily scratched). They may not absorb far-IR radiation as strongly as glass does.

Some of the main manufacturers of rigid and semi-rigid plastic sheets are:

**Kalwall Corp.** *Sun-Lite,* a flexible translucent sheet containing fiberglass and polyester. Typical thicknesses: 0.040 and 0.060 in. The material becomes slightly more cloudy (lower transmittance) if exposed for a long time to solar radiation and moisture while at high temperature. A grade called *Premium* is more durable than the regular grade.

**Filon division, Vistron Corp.** Flexible translucent sheet much like Sun-Lite except that a coating of Tedlar is included and reduces the rate of deterioration when exposed to intense sunlight in a humid, high-temperature environment.

**duPont Co.** *Lucite.* A thick transparent plate of polymethyl methacrylate.

**Rohm and Haas** *Plexiglas.* A thick transparent sheet of polymethyl methacrylate. Also, polycarbonate sheet.

**CY/RO Industries** Polycarbonate sheet. Also, Acrylic sheet.

**General Electric Co.** *Lexan.* A thick near-transparent sheet of polycarbonate.

## MANUFACTURERS OF FLEXIBLE PLASTIC SHEETS

Highly flexible (very thin—0.001 to 0.020 in.) plastic sheets have these attractive features: they are very cheap, they can be rolled up compactly for shipment, they can be rolled up on (or unrolled from) a window-top roller each morning or evening. they can be cut to size with scissors, and they can be attached and sealed with tape, tacks, staples, etc.

Some of the main manufacturers are:

**DuPont Co.** *Tedlar,* a 0.001 to 0.004-in. near-transparent sheet of polyvinyl fluoride which stands direct sunlight well and shrinks slightly when heated. Also *Tyvek,* a polyolefin fabric. Also *Teflon,* which is fluorinated ethylene propylene.

**Westlake Plastics Co.** *Kynar,* a thin sheet of polyvinylidene fluoride.

**Martin Processing, Inc.** *Llumar,* a thin polyester film.

**Warp Bros.** *Poly-Pane,* a thin transparent sheet.

**W. J. Dennis & Co.** Transparent Plastic No. 2401, which is 0.001-in. thick; part of Storm Window Kit.

**Plaskolite, Inc.** *Weatherizer,* a transparent mylar sheet.

**Technology Development Corp.** *Weatherguard,* a transparent sheet of vinyl plastic.

**Monsanto Chemical Co.** *Monsanto Film #602,* a 0.006-in. polyethylene film for greenhouse roof, etc.

**Dow-Corning**  *Silicone film.*

**3M Co.**  *Flexigard 7140; Polyethylene teraphthalate (PET).* Withstands prolonged exposure to direct sunlight. Has high reflectance; one sheet reflects 12%. (S-85i).

**Madico, Inc.** Various semi-transparent, metalized, reflecting sheets, including *Fade-Shield, Reflecto-Shade, Shade-Shield.*

**Temp-Rite, Inc.** Vinyl sheets that are transparent and (because of inherent "memory") tend to remain taut and flat.

For further information on trade names of plastic sheets and manufacturers of plastic sheets, see Appendices 6 and 7.

## MANUFACTURERS OF BUBBLE-PLASTIC AND HONEYCOMB-PLASTIC SHEETS

Because bubble-plastic or honeycomb-plastic sheets trap air, they may have much higher R-value than a simple thin sheet. Yet the cost remains very low. The sketches show front views and cross sections.

Sealed Air Corp. sells a material called Air Cap SC 120 which contains circular, close-packed "bubbles" 0.4 in. in diameter and 1/8 in. thick. In 1979 the cost was 6¢/ft² if 2000 ft² were purchased at one time.

Another producer sells a material consisting mainly of bubbles about 1 in. in diameter.

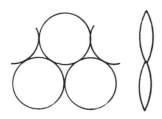

Three companies, Alpha Associates, Econ Co., and Northeast Energy Corp., are involved in the manufacture and marketing of a material consisting mainly of hexagonal bubbles of plastic. It is called Thermalite. The bubbles are 7/16 in. wide and about 1/4 in. thick. The material has a pressure-sensitive adhesive backing.

## MANUFACTURERS OF DUAL-WALL TRANSLUCENT PLASTIC SHEETS

The two planar sheets of this assembly are joined by parallel partitions. The air trapped between the partitions provides high R-value.

Some of the main manufacturers are:

**Rohm and Haas**  Makes *Tuffak Twindow* of polycarbonate. The overall thickness is about 1/4 in.

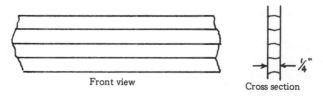

Front view                Cross section

**CY/RO Industries** Makes dual-wall sheets of polycarbonate and also dual-wall sheets of acrylic. Thickness: about 1/2 in.

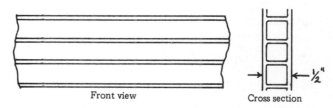

Front view                Cross section

**General Electric Co.** Makes a dual-wall sheet of polycarbonate.

**Primax Plastics Corp.** Makes polypropylene plate consisting of a corrugated sheet sandwiched between two flat sheets.

## MEANS OF ATTACHING A RIGID SHEET

There are many ways of attaching a rigid glazing sheet (of glass or plastic, transparent or translucent) to the outside of a window:

**Attach the sheet to an unmodified sash** If the rigid sheet is of plastic, one can drill holes in it and attach it to the sash by means of screws. But if the screws are situated far apart and the sheet is thin, leakage of air and water may occur at intervening locations.

A better procedure is to use a rigid retaining strip of some sort, e.g., a batten, clamping strip, framing strip, or channel. It may be wood, plastic, or metal. It may be secured to the sash frame by screws, adhesive, or tape. Sealant may be used.

If the sheet is of glass, the use of battens, clamping strips, framing strips, or channels is mandatory.

An interesting type of channel, called Solipane, is made of PVC by Temp-Rite Inc. It consists of a rigid member and flexible member, the flexibility of the latter being sufficient to accommodate plate thicknesses up to 1/4 in.

Econ Co. makes several kinds of framing or sealing strips, including kinds called Insulite and Magnetite. Dayton Corp. sells various plastic framing channels and strips (see Chapter 12).

Various kinds of channels are sold also by Plaskolite, Inc.

If the window is a conventional double-hung, two-sash type and if the additional glazing sheet is attached just to the upper sash, the room occupants can still open the window in the normal way. A good piece of advice, to be found on p. 234 of a recent book by Wade (W-40), is to attach the added glazing to the outside of the upper (outer) sash and to the inside of the lower (inner) sash. Then one can open either sash in the usual way. The added glazing sheets do not obstruct such motion.

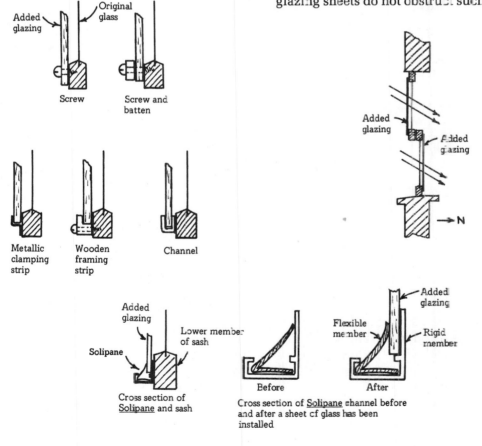

Screw

Screw and batten

Metallic clamping strip

Wooden framing strip

Channel

Added glazing

Added glazing

N

Added glazing

Lower member of sash

Solipane

Cross section of Solipane and sash

Before

After

Flexible member

Added glazing

Rigid member

Cross section of Solipane channel before and after a sheet of glass has been installed

Several companies sell framed sheets of glass for use as added glazing. Frames are of plastic or aluminum. They are attached by swiveling spring clips, or other means.

Some of the pertinent companies are:

Alcoa Building Products Inc., a subsidiary of ALCOA, Pittsburgh, PA 15219.

Thermal Units, Inc., 181 Conant St., Pawtucket, RI 02860.

**Attach the sheet to the fixed frame of the window** Any of the attachment means described above may be used. Or hinges may be used.

## MEANS OF ATTACHING A FLEXIBLE SHEET

An added sheet of flexible-plastic glazing may be attached in any of several ways described below.

**Attach the sheet directly to the glass pane, with no intervening airspace** Use tape or cement. Or use a commercially available kind of insulating plastic that has a factory-applied sticky coating on one face. Example: Thermolite, made by Econ Co. This sheet, consisting of a close-packed array of hexagonal plastic bubbles 1/2-in. wide and 1/8-in. thick, has a sticky coating (pressure-sensitive adhesive) on one face.

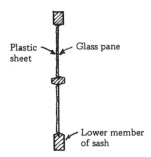

**Attach the sheet to the glass pane but interpose spacers that provide a region of trapped air** This entails a little more work, but it cuts the heat-loss more. The spacer may consist of slender strips of wood, cardboard, or plastic—or thin tubes, e.g., transparent plastic drinking straws. Moisture build-up may occur.

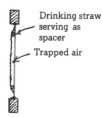

**Attach the sheet to the sash** This is easily done. The trapped air helps cut heat-loss. But moisture build-up may occur.

The plastic sheet may be attached as is by means of tape. Or it may be attached to a frame which is then fastened to the sash by screws, tape, or channels. If channels are used, frame-and-sheet can be removed (slid out) quickly at any time, to permit cleaning the window or to allow moisture to evaporate. The frame itself may be of wood, plastic, or metal. It can be removed when spring comes.

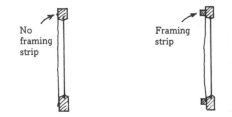

If *both* sides of the frame are covered with plastic, heat-loss is cut even further, at little extra cost.

The frame may consist of separate pieces (battens) each of which is nailed, taped, or glued to the sash. Wood, plastic, cardboard, or metal may be used. Construction is simple, but removing and reinstalling the equipment is time-consuming. Butt joints or mitred joints may be used.

Or the frame may consist of special, mass-produced locking strips. Each strip grasps and

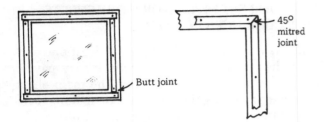

Butt joint

45° mitred joint

holds one edge of the plastic sheet and is affixed to the sash by nails, screws, or double-faced tape. For example, use PVC locking strips produced by Temp-Rite Inc. Cross sections of some of these strips are shown in the accompanying diagrams.

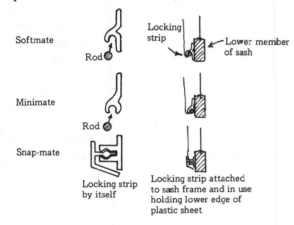

Softmate

Rod

Minimate

Rod

Snap-mate

Locking strip by itself

Locking strip

Lower member of sash

Locking strip attached to sash frame and in use holding lower edge of plastic sheet

Softmate, for example, consists of a PVC main strip and a PVC locking rod. After the flexible sheet has been laid across the mouth of the strip, the locking rod is forced into place (in the groove, or slot) to hold the sheet securely.

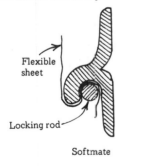

Flexible sheet

Locking rod

Softmate

The Weatherguard system, developed by Technology Development Corp. of Boston, Mass., employs a 0.012-in. sheet of vinyl plastic (called Vinylglass) that includes an inhibitor for increasing the resistance to UV radiation. The sheet is held within a frame of rolled aluminum. Nylon pile weatherstripping is included. Special fastening devices are provided. On both sashes (upper and lower) the additional glazing is on the outdoor side. The sheet attached to the lower sash can be released (via a mechanism that passes through the sash frame) from indoors.

**Attach the sheet to the fixed frame of the window**
Here any of the above-described attachment means may be used.

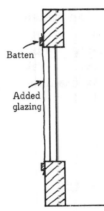

Batten

Added glazing

## WHICH ATTACHMENT LOCATION IS BEST?

If the additional glazing is attached directly to the glass panes, it is sure to reduce heat-loss through the panes. The reduction is accomplished irrespective of any cracks around the sashes or cracks in the fixed frame of the window.

If it is attached to the frame of the sash, there is a risk as well as some benefit. The risk is that, if there are gaps at the edges of the added glazing, much cold air may enter the space between this sheet and the original glass panes—with the consequence that loss of heat through the original panes is great. The benefits are: (1) if the additional glazing is well sealed, it prevents in-leak of cold air via the cracks around the original panes, and (2) the region of trapped air helps reduce heat-loss.

If the additional glazing is attached to the fixed frame of the window, there is again the risk that if the edge seals are very poor, heat-loss will remain great. However, if the seals are good, then in-leak of cold air via cracks at edges of the sashes and via cracks in the fixed frame of the window is stopped. Note that attaching the additional glazing to the fixed frame of the window entails use of a larger sheet, and this may require that the glazing be thicker. Also, a larger sheet is more prone to flap or belly in the wind and more prone to hurt the appearance of the house.

## HOW TO PREVENT A LARGE OUTDOOR SHEET OF FLEXIBLE PLASTIC FROM FLAPPING AND BELLYING

To prevent flapping and bellying, install some intermediate bars (vertical or horizontal) placed so as to urge the plastic outward or inward, thus increasing the tension on it, i.e., taking up the slack. The set of bars may, for example, impose a slight "crown" on the flexible sheet.

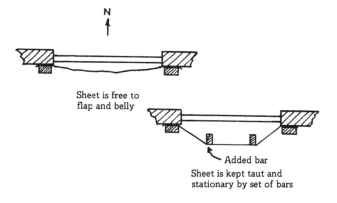

N

Sheet is free to flap and belly

Added bar
Sheet is kept taut and stationary by set of bars

Horizontal cross section of large window that has been equipped with a large outdoor sheet of flexible plastic

Where very large areas are covered with two sheets of plastic (double glazing of plastic), flapping may be prevented by employing a very small blower to maintain a little air pressure in the space between the sheets. If 0.006-in. Monsanto polyethylene film #602 is used, it is recommended that a 30-watt blower be used for each 10,000 ft. of area.

Blower

Two sheets of flexible plastic held taut by air pressure

## OUTDOOR SET OF TWO OR MORE SHEETS OF TRANSLUCENT GLAZING

If one is committed to installing at least one outdoor sheet of translucent glazing, it may pay to install, in fact, two such sheets. The main effort comes in installing the outer sheet—making it resistant to wind, rain, and snow. Only a little additional effort is needed to install a second sheet—close behind (and protected by) the first. The second sheet can be of very cheap material: material that is thin, limp, and weak.

To install a thin sheet is very easy. Almost any informal installation scheme is adequate. The sheet may be attached to the window proper, or to the fixed frame of the window—or to the north side of the outer sheet itself (in which case an intervening airspace may be provided by suitable spacers, or perhaps just by the natural waviness of the inner sheet). If a special frame is provided, one plastic sheet may be attached to one face of the frame and the other to the other face.

For example: Make the outer sheet of Kalwall Sun-Lite and the inner sheet of bubble plastic.

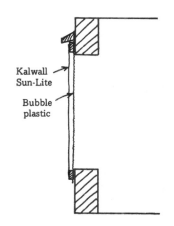

Kalwall Sun-Lite

Bubble plastic

# CHAPTER 7

# OUTDOOR OPAQUE SHUTTERS THAT ARE NOT PERMANENTLY ATTACHED

- Main Design Features
- Outdoor Single Plate Held By Buttons (Scheme 7.1)

Here I discuss outdoor shutters that consist of opaque insulating plates and are not permanently attached, i.e., shutters that are removed each morning by hand and are reinstalled each evening.

A favorable feature is that the shutter (the plate) does not have to be very strong. It can be of simple, lightweight construction. By contrast, a shutter that is to be permanently attached by hinges must be strong enough so the hinges cannot pull away and so that the shutter will not be distorted, when partially open, by its own weight or by high winds.

The following sections discuss the main design features and several complete designs.

## MAIN DESIGN FEATURES

### Shutter Proper

One may use a simple 1-inch-thick plate of Thermax, made by Celotex Corp., or High-R Sheathing, distributed by Owens-Corning Fiberglas Corp. The two products are practically identical. Both have aluminum-foil faces. The R-value of a 1-in. plate is about 8. The material is easily cut to size with a saw or sharp knife, and a waterproof tape may be applied to the edges. The material is so strong that, for a small window, a plate only 1/2-in. thick may be strong enough. For a very large window, a 1½-inch-thick plate may be appropriate. If the bare aluminum faces seem unattractive, they may be painted with almost any kind of paint.

Alternatively, one may use Styrofoam SM plates. Probably such plates should be protected by a coat of paint. Only latex paint should be used.

If some small portions of the shutter are to be subject to wear, for example edge regions engaged by buttons, they may be reinforced by taped-on pieces of sheet aluminum of the type commonly used for roof flashing, i.e., sheets 0.020 to 0.030 in. thick.

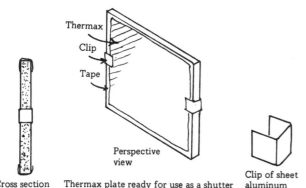

Cross section    Thermax plate ready for use as a shutter    Clip of sheet aluminum

Perspective view

Thermax
Clip
Tape

### Handles

Handles may not be needed. One can grip the shutterbyits edges.

If handles are needed, they can be made easily. One way is to extend the aluminum clips used at the edges—allow extra material and bend it to form a handle of convenient thickness. As before, the clips are secured with tape.

Another way is to use a separate strip of aluminum, bent into the shape of a handle. It may be attached with tape.

Warning: Thermax, Styrofoam, etc., consist mainly of air or other gas. They provide little grip for screws and nails. Tape, however, provides a firm grip.

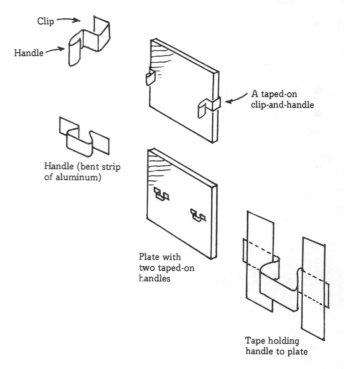

Clip
Handle

A taped-on clip-and-handle

Handle (bent strip of aluminum)

Plate with two taped-on handles

Tape holding handle to plate

### Painting the Thermax Plate

If the bare aluminum faces of the Thermax plate are deemed unattractive, paint them, or apply some attractive covering. The isocyanurate core of a Thermax plate is such an excellent insulator that painting the aluminum faces—and thus greatly reducing the reflectance with respect to far-IR radiation—does not appreciably impair the thermal performance of the plate. In tests made in March 1979 on a 1/2-in. plate of Thermax, I found that the heat-saving due to the plate was only slightly reduced when I covered both faces with green paint or black paint. Even stripping off the aluminum foils entirely decreased the heat-saving only slightly. (A sheet of aluminum foil by itself produces a large heat saving, as explained on p. 49, and much of this stems from the high reflectance with respect to far-IR radiation; but when such foil is applied to a plate of isocyanurate foam, the incremental benefit is minor.)

Another consequence of painting the plate is that the tapes (along the edges, and on the handles) become much less noticeable.

### Seals

I know of no detailed and reliable information on tolerances on gaps at the edges of outdoor shutters. Probably the tolerances depend strongly on the direction and speed of the wind. My guess is that gaps 1/32 in. or less in width are tolerable. See also Appendix 2.

If the gaps threaten to be greater than this, sealing strips should be used. Such strips may be of various shapes, materials, etc.

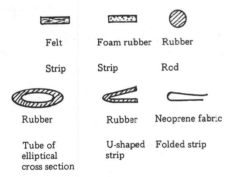

| | | |
|---|---|---|
| Felt | Foam rubber | Rubber |
| Strip | Strip | Rod |
| Rubber | Rubber | Neoprene fabric |
| Tube of elliptical cross section | U-shaped strip | Folded strip |

Should the sealing strips be attached to the fixed window system or to the (removable) shutter? To the shutter, preferably. The strips can be attached to the shutter in the workroom where the shutter proper is made. When the shutter is stored (in a garage, say) for the summer, the strips too are in a cool, dry, shaded place. When the shutters are in storage the window systems have their normal appearance: no sealing strips are present.

### Strengthening the Plate

If the shutter is to be subjected to rough handling, strengthening the plate may be desirable. The usual procedure is to construct a box-like enclosure. A rectangular wooden frame is used and face sheets are attached to it. The core may consist of Styrofoam SM or other foam. Usually the frame is made of wood. The faces may be of Masonite, or outdoor-type plywood, or hardboard, or other tough sheet. The components may be screwed, nailed, stapled, or cemented together. Edges and cracks may be painted to exclude rain.

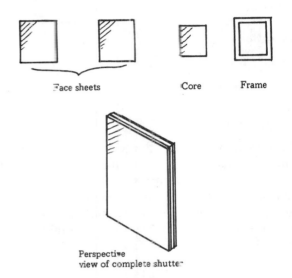

Face sheets      Core      Frame

Perspective view of complete shutter

### OUTDOOR SINGLE PLATE HELD BY BUTTONS (Scheme 7.1)

This shutter consists of little more than a single sheet of 1-inch-thick Thermax or the equivalent held in place by four buttons. No frame or stiff edging is used. The edges of the plate are covered by tape and the faces of the plate may be painted to improve their appearance. If reasonably well sealed, the shutter provides an R-value of about 8, or slightly higher if the plate traps a region of air 1/2 in. or more in thickness.

**Handles and seals**    These have been discussed on a previous page.

**Cost**    1-in. Thermax costs about $0.50/ft², retail.

**Storage**    When not in use, the shutter may be set on the ground, e.g., behind a bush or post or fence that will keep it from being blown away in a high wind.

**Limitations**   The plate may be damaged by rough handling and perhaps by woodpeckers, squirrels, etc. Rain may penetrate behind it unless it is protected by eaves.

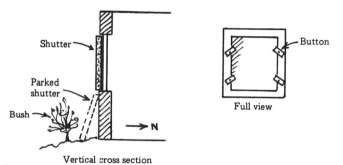

Vertical cross section

Full view

### Scheme 7.1a

As above, except omit the two upper buttons and employ, instead, a horizontal cap strip (with beveled top and bottom) secured and sealed to the top of the fixed frame of the window. This strip holds the upper edge of the shutter captive and also prevents rainwater from dribbling down behind it.

### Scheme 7.1b

As above, except cut a small central hole in the shutter with a pocket knife or small saw and cover the hole, on both faces, with transparent plastic secured with waterproof tape. Now, the shutter can be left in place all day as well as all night (unless it is a south window, or other window that might receive much solar radiation or is of importance for illumination or view). Enough daylight can enter, via the hole; to illuminate the room barely adequately; yet the capability of the shutter to reduce heat-loss is scarcely impaired at all.

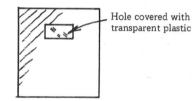

### Scheme 7.1c

As above, except use an unpainted Thermax plate and, when parking it, secure it to a special, near-horizontal stand near the base of the window so that it will reflect much solar radiation toward the window during the middle of the day.

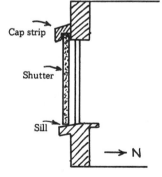

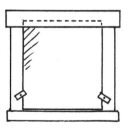

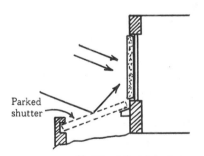

Aluminum-faced shutter can serve as a reflector

# CHAPTER 8

# OUTDOOR OPAQUE SHUTTERS ATTACHED BY HINGES

- Outdoor Single Plate Attached by Hinges at One Side (Scheme 8.1)
- Outdoor Shutter Employing Two Plates Hinged at the Sides (Scheme 8.2)
- Outdoor Shutter Employing Two pairs of Hinged Folding Plates Operable from Indoors (Thermafold Shutter) (Scheme 8.3)
- Outdoor Single Plate Attached by Hinges at Top (Scheme 8.4)
- Outdoor Single Plate Controlled by Cord Attached to Arm (Scheme 8.5)
- Outdoor Single Plate, Hinged at Top, That Employs Four Tricks to Make Plate Close Tightly Automatically (Scheme 8.6)
- Outdoor Single Plate, Hinged at Top, Raised with the Aid of a Pulley Attached to Eaves (Scheme 8.7)
- Outdoor Single Plate That Can Be Secured at the Bottom by a Latch System Controlled by the Same Cord Used in Raising the Plate (Scheme 8.8)
- Outdoor Single-Plate Hinged at Top and Controlled by Worm-and-Gear (Saskatchewan Outdoor Shutter) (Scheme 8.9)
- Outdoor Single Reflective Plate Hinged at Bottom (Baer Shutter) (Scheme 8.10)
- Outdoor Single Reflective Plate Hinged at Bottom in Winter and at Top in Summer (Scheme 8.11)

Here I describe shutters that are permanently attached by hinges.

What kind of plate should be used? Wood? Weatherproof plywood? Rigid foam edged with wooden strips and faced with outdoor-type plywood? For sure, if a shutter is to be mounted outdoors and attached by hinges, it must be made strong enough so that the hinges will not tear loose while the shutter is being opened or closed on a very windy day.

What kind of seals should be used? See Chapter 4 and Appendix 2. The seals may be attached to the shutter or to the fixed frame of the window.

Where should the hinges be attached? At the side, top, or bottom? At the side, usually, because then there is no laborious raising or lowering necessary. Also there is nothing that can swing downward vigorously and possibly strike someone. The accompanying sketches show an outdoor shutter that is hinged at the west side.

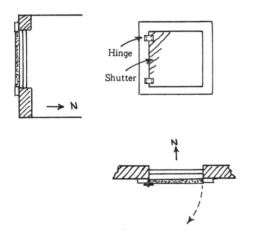

**What kinds of hinges should be used?**   The hinges should be of non-corroding material such as brass or galvanized steel. They should be strong enough to hold steady even when high wind strikes the shutter. The screws used should be long enough to penetrate deep into the wood comprising the fixed frame of the window; that is, they should reach well beyond the outermost region which may deteriorate after a few years of exposure to sun and rain.

**Possible need for shims at hinges**  Consider an outdoor shutter that is hinged at one side and is serving a south window. It may happen that such a

shutter is installed incorrectly: the hinged edge does not lie in the desired plane. If this edge is a little too far north, the compression-type sealing strip may be excessively compressed when the resident tries to close the shutter; the shutter may fail to close fully and cold outdoor air may circulate between shutter and window. If the hinged edge of the shutter is a little too far south, the sealing strip there may fail to make full contact and thus may fail to do its job.

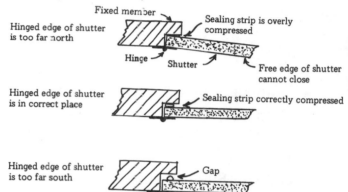

Horizontal cross section of shutter, hinge, and sealing strip

What is to be done if the hinged edge of the shutter is too far in or too far out? The easiest solution is to install a shim of just the right thickness to correct the situation, i.e., to alter the plane of the (closed) shutter so that the sealing strip is compressed by about 20%. See accompanying sketches. The shim may be of wood, neoprene, or other durable material.

Shutter shimmed outward

Shutter shimmed inward

The same general problem may arise with a shutter hinged at the top or bottom, and the same solution is possible: shim the hinged edge of the shutter in or out, as required.

**Use of a rain-deflector strip**   It may pay to install, just above the window, an outdoor horizontal strip that will intercept rain (rain drops falling and rainwater running down the face of the house) that might otherwise find its way in behind the shutter. The deflector-strip should be undercut to insure that all water that drips from it falls well clear of the shutter.

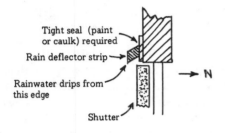

## OUTDOOR SINGLE PLATE ATTACHED BY HINGES AT ONE SIDE (Scheme 8.1)

Such shutter is sketched in the previous section.

The shutter (with sealing strips) may press directly against the glass or—more likely—may press against the frame members of the sash or against the fixed frame (housing) of the window as a whole.

When the shutter is closed, it may be held tightly closed by means of an outdoor, manually operated button or hook.

When the shutter is open, it may be held by another button or hook. It must be held snugly, otherwise it will shift about and bang in a gusty wind.

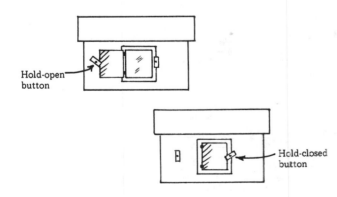

## OUTDOOR SHUTTER EMPLOYING TWO PLATES HINGED AT THE SIDES (Scheme 8.2)

Here, each plate is of half-width, hence is lighter in weight and easier to swing. Each half has its own hold-open button. A single hold-closed button—at bottom center—may suffice.

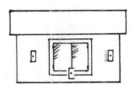

Tighter seal is achieved by use of two hold-closed buttons: one at the top and one at the bottom. If the upper one is too high to reach, use a latching device that is closed by spring pressure and is opened by pulling down on a long string or chain.

Another way to secure the top of a high plate is to mount the button on a vertical batten that is affixed to one plate and slightly overlaps the other (to cover the gap and keep rain out). The button engages a restrainer arm fastened to the fixed frame of the window. Attach long cords to the ends of the button; by pulling down on the appropriate cord you can lock or unlock the button.

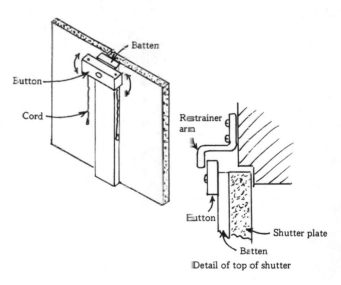

Detail of top of shutter

## OUTDOOR SHUTTER EMPLOYING TWO PAIRS OF HINGED FOLDING PLATES OPERABLE FROM INDOORS (THERMAFOLD SHUTTER) (Scheme 8.3)

The Thermafold shutter, made and sold by Shutters, Inc., was developed by P. W. Swanstrom (see Patent 4,044,812 of 8/30/77). It is intended mainly for use with sliding glass doors, but presumably can be applied to windows also.

There are two assemblies: a pair of panels (plates) for the east side of the glass-door area and a pair for the west side. Each pair includes a narrow plate, attached to the wall of the house by hinges, and a wide plate attached by hinges to the narrow plate. Each plate is 1¼ in. thick; it consists of a 1-in. urethane foam sheet faced on both sides with 1/8-in. sheets of tempered hardboard. Each edge consists of a wooden strip. Each hinged edge of each panel is served by three hinges. The outer edges of the wide plates are not free but are constrained by small nylon sliders at top and bottom; these slide in fixed horizontal weatherstripped channels of anodized aluminum.

Each pair of panels is opened and closed from within the house by means of a crank and a worm-and-gear system.

The equipment may be used in summer to exclude solar radiation.

Cost: About $260 F.O.B., not including installation. Installation by professionals adds about $120 to the cost.

## OUTDOOR SINGLE PLATE ATTACHED BY HINGES AT TOP (Scheme 8.4)

The single rigid insulating plate (with foam core, strong frame, and sealing strips, all as described in the previous chapter) is attached at the top by hinges. When open, the shutter is held up by a long pole; the upper end of the pole is attached to the edge of the shutter by a short cord. The lower end of the pole is stuck an inch or two into the ground. When the shutter is closed, it is secured by a button.

Note: When the shutter is up it lies close beneath the eaves, which project several feet and protect the shutter from snow, rain, and wind.

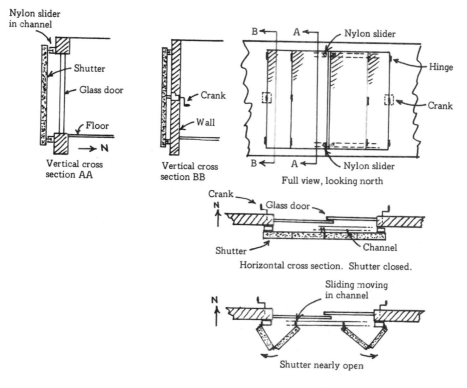

Vertical cross section AA

Vertical cross section BB

Full view, looking north

Horizontal cross section. Shutter closed.

Shutter nearly open

In summer the shutter can be left closed to exclude solar radiation. Or it can be closed part way to serve as a canopy or awning.

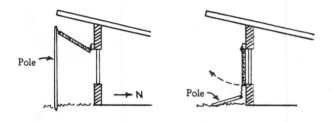

## Limitations

Use of a pole is awkward in several respects. For example, opening the shutter (raising it) with the aid of the pole is difficult, especially if the shutter is very heavy.

When the shutter is open, a gust of wind may lift it higher and knock the pole askew, allowing the shutter to fall.

### Scheme 8.4a

As above, except hoist the shutter with the aid of a cord that runs over a pulley mounted just beneath the eaves and provide a lock-on hook that holds the shutter up. The shutter cannot be lowered until the hook is unlocked (by a pull on a release-string, if the hook is too high to reach).

It may pay to provide a stop for the upward swing of the shutter. Thus when the shutter is fully raised and hooked, there is no vertical play: the wind cannot make the shutter move.

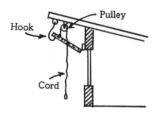

## OUTDOOR SINGLE PLATE CONTROLLED BY CORD ATTACHED TO ARM (Scheme 8.5)

Here a rigid arm, extending 1½ ft. outward from the shutter and attached firmly to it by means

of screws, controls the motion of the shutter. Attached to the outer end of the arm is a cord which runs over a pulley, 1½ ft. above the top of the window, into the house.

To open the shutter the occupant of the room pulls on the cord. He continues to pull until the shutter can open no farther—because the arm has made contact with the face of the house. He then secures the cord. Because the cord is snug, the shutter has no free play; it does not move even when hit by gusty winds.

To close the shutter, the occupant releases the cord gradually, allowing the shutter to descend slowly.

The shutter will close tightly enough of its own accord if:

the hinges are nearly frictionless,

the sealing strips are engaged only very lightly,

the arm is heavy enough to exert an appreciable gravitational torque, and

there are no strong winds.

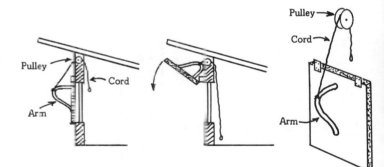

## Limitations

If the operator is careless in lowering the shutter, he may allow it to fall too fast. However, during the last portion of its downward swing it will be slowed (damped) by trapped air.

Because the closed shutter is not locked closed, it may be banged somewhat by a gusty wind. (One could install a manually operated button to secure the lower edge of the shutter. But the operator would have to go outdoors to turn the button.)

*Scheme 8.5a*

As above except:

1. Make the shutter extremely light, so that raising it is easy and if it falls it will fall slowly (thanks to wind resistance). Make it of Thermax or other rigid foam.

2. Use two pulleys in tandem: one on either side of the 8-inch-thick wall. This obviates the need to mount a (large) pulley within the wall.

3. Provide a clamping device that will hold the shutter firmly shut and can be operated from indoors. For example, provide an outdoor hook attached to a horizontal rod that runs through the wall of the house into the room and is urged northward by a helical spring. To clamp the shutter, the operator pushes the rod southward an inch or two, turns it 180 degrees (so that the hook aims upward), and releases it (so that the spring drives the rod northward, pressing the hook against the lower edge of the shutter). When the shutter is to be opened, the operator pushes the rod southward, turns it 180 degrees, and releases it; the shutter is then free to be raised.

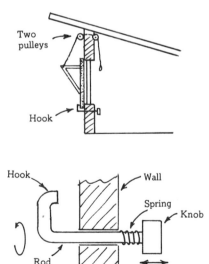

*Scheme 8.5b*

As above, but include a counterweight, attached to the indoor end of the cord, which will facilitate raising or lowering the shutter (by canceling its weight). The counterweight consists of a block of wood, or (safer) a sand-filled bag.

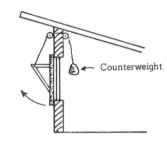

## OUTDOOR SINGLE PLATE, HINGED AT TOP, THAT EMPLOYS FOUR TRICKS TO MAKE PLATE CLOSE TIGHTLY AUTOMATICALLY (Scheme 8.6)

An outdoor, south-facing shutter that is hinged at the top in a simple naive way, i.e. with the hinge-pins approximately in the plane of the outer face of the shutter, tends to hang slightly open because the center of gravity lies somewhat to the north of the vertical plane that includes the hinges. Because the pull of gravity keeps the shutter slightly open, outdoor air can circulate between shutter and glass.

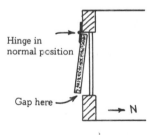

There are four tricks that may be used—individually or in combination—to make such a shutter close completely automatically.

1. Place the hinges farther inward, relative to the vertical midplane of the shutter, so that the vertical plane that includes the hinge-pins is farther inward than the midplane of the (closed) shutter.

Warning: If the hinges are set far inward, precautions must be taken to insure that there will

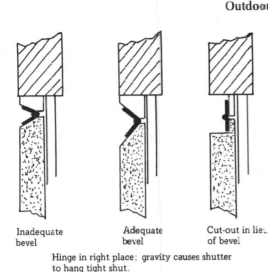

Inadequate
bevel

Adequate
bevel

Cut-out in lieu
of bevel

Hinge in wrong place;
gravity causes shutter
to hang partly open.

Hinge in right place; gravity causes shutter
to hang tight shut.

be no interference when one attempts to open the shutter. For example one may bevel the surfaces that threaten to interfere.

When the hinge is correctly placed, gravity tends to close the shutter fully. Even when the shutter is closed, there is a residual positive closing torque. And if the extent of bevel is adequate, the shutter can be opened at least 90 or 100 degrees.

Two alternative hinge schemes are:

Attach the hinges directly to the glass (suggested by S. C. Baer).

Use pivot-type hinges, with the pivots attached to the sides (jambs) of the fixed window frame.

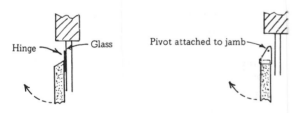

Hinge

Glass

Pivot attached to jamb

2. Use spring-biased hinges, i.e., hinges that include coil springs designed to exert a closing torque even when the shutter is closed. Such hinges are commonly used to keep screen doors shut. One could design a special spring that would maintain a very large torque—and the spring could be merely a spring, i.e., not a hinge also.

3. Attach, to a 6-inch-long arm that projects southward from the shutter, a 10-lb. weight. The pull of gravity on this weight will produce a positive closing torque at all times—even when the shutter is closed. Of course, the presence of the weight will make the shutter harder to open, but if the weight is mounted high up the added difficulty is negligible.

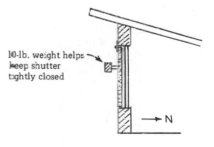

10-lb. weight helps
keep shutter
tightly closed

N

4. Cement some small permanent magnets to the lower portion of the shutter and cement a corresponding set of cold-rolled steel pieces to the fixed frame of the window When the shutter is within 1/16 in. of fully closed position, the magnets produce a strong closing force. Alternatively, use magnetic tape such as is used on refrigerator doors.

Using such tricks, the homeowner may find that—if his house is in a region where there are many trees and shrubs to block the wind—the

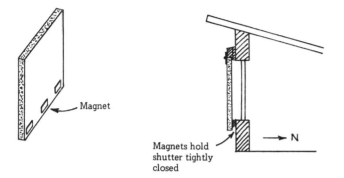

Magnet

Magnets hold
shutter tightly
closed

→ N

shutters remain tightly enough closed so that no
clamps or buttons are needed. If violent winds
sometimes occur, a more secure closure may be
needed.

### OUTDOOR SINGLE PLATE, HINGED AT TOP, RAISED WITH THE AID OF A PULLEY ATTACHED TO EAVES (Scheme 8.7)

Here, the main pulley is attached to the underside
of the eaves and is about 2 ft. from the south wall.
This provides better mechanical advantage in
starting to open the shutter.

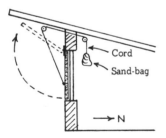

Cord

Sand-bag

→ N

### OUTDOOR SINGLE PLATE THAT CAN BE SECURED AT THE BOTTOM BY A LATCH SYSTEM CONTROLLED BY THE SAME CORD USED IN RAISING THE PLATE (Scheme 8.8)

The special feature of this shutter, which has
hinges at the top and is controlled from indoors by
means of a cord running over pulleys, is the fully-
automatic latch system at the lower edge of the
shutter. A single motion of a single cord lowers
and latches the shutter, and a single counter-
motion releases the latch and raises the shutter.

The latch employs a horizontal bar situated
near the lower edge of the shutter and permanently
(but loosely) attached to it by two yokes which
allow the bar 3 inches of vertical free play. The
outdoor end of the above-mentioned rope is tied to
the midpoint of the bar. The bar is 4 in. longer than
the shutter is wide; thus the bar projects 2 in. at
each side.

Affixed to the wall of the house are two sturdy
keepers. When the shutter is closed and the cord is
then relaxed, the bar drops down about 2 in. and
each end engages a keeper. The keeper notches are
tapered in such a way that the small weight of the
bar produces a 2-times greater force urging the
shutter northward, i.e., farther closed. If a gusty
wind jiggles the shutter, the bar sinks deeper into
the notches, making the closure even tighter.
(Note: when the counterweight is lifted, it is raised
an inch or two "extra far" so as to relax the cord; the
counterweight rests on a small shelf provided, or is
secured to a horizontal peg provided, so that the
cord will remain relaxed.)

When the cord is pulled (by pulling down on
the counterweight) the outdoor end forces the bar
upward, free of the keepers. Further pulling on the
cord raises the shutter (because the bar is attached
to it by the yokes).

In summary, the shutter is opened, closed, and
latched at two lower corners entirely from within
the room and solely by operation of one cord.

### Comment

The shutter can be used in summer to exclude solar
radiation. Also it can be left partially open (at 45
degrees, say) to act as an awning to exclude direct
solar radiation but admit light and air. An additional device is needed to lock the shutter in the
45-degree position.

### Limitations

The general complexity of the device may be a
drawback.

The shutter, when open, projects outward
quite far. Unless the eaves are very wide, the
shutter will project and intercept rain, snow,
and wind. (Making the eaves very wide would

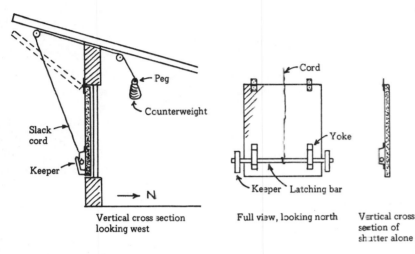

Vertical cross section
looking west

Full view, looking north

Vertical cross
section of
shutter alone

solve the problem and also would help exclude solar radiation in summer.)

If the shutter is lowered rapidly, it could strike
someone (standing outdoors, near the window) on the head and hurt him.

The latching bar may fail to engage the keepers
unless the overall design is such that the shutter tends to hang almost vertically downward
even where is a little upward-to-south tension
on the haul-up cord. Use of spring-biased
hinges may be advisable.

## OUTDOOR SINGLE PLATE HINGED AT TOP AND CONTROLLED BY WORM-AND-GEAR (SASKATCHEWAN OUTDOOR SHUTTER) (Scheme 8.9)

Use of a hinged, swing-up plate of Styrofoam was
proposed, for application to a 6 ft. × 4 ft. south
window in Canada, early in 1977 by the designers
of Saskatchewan Energy Conserving Demonstration House. Sponsor of house: Saskatchewan
Housing Corp. Chief architect: Ken Scherle. The
4-inch-thick, 6 ft. × 4 ft. plate is faced with an
aluminum sheet and has strong edge pieces. It
contains internal stiffeners consisting of a 1½-in.
aluminum "angle-iron." There are two hinges at
the top. When the shutter is closed, by means of a
crank at shoulder height, it presses against tubular
rubber seals.

A short metallic arm is affixed to one upper
corner of the shutter, and this can be turned—so as
to open or close the shutter— by means of a worm-
and-gear system that is attached to the outer face of
the wall. The worm gear is turned manually by
means of an indoor crank, the shaft of which passes
through the wall. On sunny days in winter the
shutter is swung upward through an angle of about
100 deg. and is then almost touching the roof overhang, which protects it from rain, snow, and wind.

On hot summer days the shutter is kept at 60
deg. from the vertical and acts as an awning that
admits much daylight but excludes direct solar
radiation.

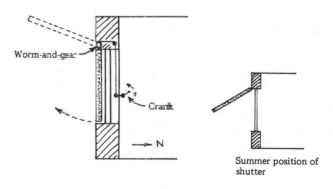

Summer position of
shutter

### Limitations

The shutter is complicated and expensive.

Because it is operated and supported from one corner only, it must be very strong and warp resistant.

The worm-and-gear system, required to withstand large snow loads and wind loads, must be very strong and must have little free play.

## OUTDOOR SINGLE REFLECTIVE PLATE HINGED AT BOTTOM (BAER SHUTTER) (Scheme 8.10)

This shutter was developed about 1970 by S. C. Baer. Installed on his house in Corrales, New Mexico, it has performed excellently.

Each face of the 3-inch-thick shutter consists of a 0.024-in. sheet of satin-anodized aluminum. Sandwiched between these is a strong honeycomb structure filled with urethane foam.

The shutter is attached outdoors at the base of a very large window. Three hinges are used, and each is of special, offset-and-underslung type that allows room for a sealing strip and makes all of the screws easily accessible. When the shutter is open, its free end rests informally on the ground; it is strong enough so that people are able to walk across it. High winds do not disturb it. When it is closed, its all-around sealing strip (a folded strip of neoprene fabric) is compressed between the shutter and the fixed wooden frame of the window.

The shutter is closed (raised) by means of a 3/8-inch-diameter nylon rope that is attached to the top center of the shutter and runs informally through a small hole above the center of the window. Friction is reduced by an idler pulley at the outdoor end of the hole. Indoors, the rope is attached to a manually operated winch such as is used for pulling small boats onto trailers. There is no latch or lock; the shutter is kept firmly closed by the tension and elasticity of the nylon rope. (When the shutter is almost completely closed, high wind may make it move about; but the motion is smooth because of the damping effect of air trapped between shutter and window.) If, when the shutter is partly closed, the crank-arm of the winch is released, the shutter will fall; but this has not happened. Sometimes, when the shutter has been closed, special steps are taken, e.g., with the aid of a chain, to make it almost impossible for small children to open it.

The three functions of the shutter: (1) The most important use is in excluding solar radiation in summer: the shutter is kept closed most of the summer. (2) Next in importance is serving as a reflector (when open, on sunny winter days) to direct more solar radiation to the large window. (3) Heat-loss is reduced on winter nights with the shutter closed. The shutter is opened and closed daily in winter only.

### Comment

The shutter is very strong. It has never been damaged or warped. It has performed its three functions well. In summer (with the shutter kept closed), the owner grows tall plants closely adjacent to it; that is, there is no waste of growing space.

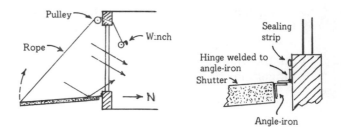

### Limitations

When the shutter is closed there is a danger to anyone directly outside the window if someone inadvertently lowers the shutter suddenly.

When the shutter is open, it preempts valuable space on the south side of the house.

Opening the shutter is difficult, or a little precarious, in high wind.

### Owner's Comment

In a future house, he would use smaller shutters and operate each directly, manually, from outdoors.

### Scheme 8.10a

Approximately the same as above except that the north edges of the shutters are about 18 in. above ground level, so that on sunny days in December, when the panels are open and are reflecting additional solar radiation into the rooms, the shutters can slope downward to the south (rather than being horizontal) and thus intercept a larger amount of solar radiation. Such shutters were made by Geoffrey Gerhard (G-140).

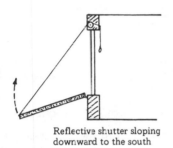

Reflective shutter sloping downward to the south

### Scheme 8.10b

This shutter is smaller and lighter. When it is open, its free end rests on two support posts and may be attached to them by means of clips, latches, or ties, so that wind cannot disturb it. On sunny days the upper face of the shutter reflects much solar radiation toward the window.

To close the shutter, the free end is raised and pushed against the fixed frame of the window, where suitable sealing strips are provided. It is secured by means of two buttons.

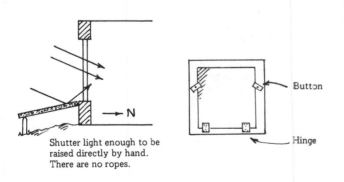

Shutter light enough to be raised directly by hand. There are no ropes.

### Scheme 8.10c

As above, but use automatic spring-actuated latches instead of buttons.

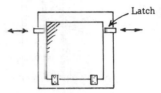

### Scheme 8.10d

Dispense with buttons and latches. Instead, provide, for each upper corner of the shutter, a small counterweight attached by a cord running over a pulley. The counterweights facilitate opening and closing the shutter and—more important—exert a fixed closing force to insure a tight seal.

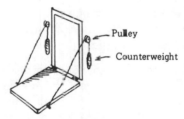

## OUTDOOR SINGLE REFLECTIVE PLATE HINGED AT BOTTOM IN WINTER AND AT TOP IN SUMMER (Scheme 8.11)

The shutter is 1 or 2 in. thick, contains foam-type insulation faced with aluminum, has a rigid, warp-resistant frame, and is attached to the fixed frame of the window by hinges.

There are two alternative hinge locations:

*In winter* the hinges are at the lower edge of the shutter. On a sunny day the shutter is open and the free end rests on adjustable-height posts (which are adjusted manually every few weeks) and much solar radiation is reflected toward the window. At dusk the shutter is closed—raised and pressed against the seals attached to the fixed frame of the window—and held shut by buttons.

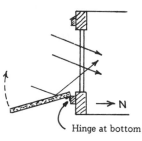

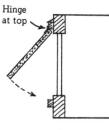

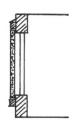

Sunny day in winter     Sunny day in summer     Nighttime in winter (or very hot day in summer)

*In summer* the hinges are at the upper edge of the shutter. On a very hot sunny day, the shutter is left closed to keep solar energy out. On moderately hot sunny days the shutter is opened 60 degrees to act as an awning (blocking direct sunlight but admitting indirect radiation and providing view). It is held at this angle by means of tie rods. On cold summer days the shutter may be raised high, so that it almost touches the eaves, to admit direct and indirect solar radiation.

To make the changeover from use of bottom hinges to use of top hinges, the resident merely removes the pins from the bottom hinges (while the shutter is held closed) and installs them in the top hinges.

**Comment**

The shutter performs certain valuable functions in winter and other valuable functions in summer.

It is simple, durable, and cheap.

It is especially effective when applied to a wide window. If the window and shutter are wide, the open or partly-open shutter continues to perform the desired functions fairly well even an hour or two before or after noon.

*Scheme 8.11a*

As above, except provide counterweights to facilitate raising and lowering the shutter.

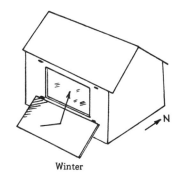

Winter

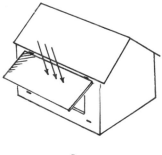

Summer

Two important daytime uses of an extra-wide reflective shutter that can be hinged at bottom or top

## Scheme 8.11a

As above, except make the shutter much wider than the window so that, when the shutter is open, it will serve its (winter or summer) daytime functions throughout a greater portion of the day.

## Scheme 8.11c

As above, except design the support system in such a way that when the plate is in use as a reflector, and then receives a heavy load of snow, the shutter will automatically tilt more steeply to allow the snow to slide off, and the shutter will then revert to the normal tilt. (Suggested by E. Moore.)

# CHAPTER 9

# OUTDOOR OPAQUE SHUTTERS THAT SLIDE TO RIGHT OR LEFT

- Outdoor Single Sliding Plate Suspended from an E-W Track (Zero Energy House Shutter) (Scheme 9.1)
- Outdoor Tandem Sliding Plates Suspended from an E-W Track (Goosebrook House Shutter) (Scheme 9.2)
- Outdoor Single Sliding Plate That Has Compliant Suspension, Slides with Little Friction, and Is Clamped Tightly Against Window Frame (Scheme 9.3)

Outdoor shutters that slide in horizontal direction are rare. They can be designed to reduce heat-loss drastically, but they can be heavy and expensive.

I describe first a design that may be the world's most elegant outdoor shutter. Then I describe some designs that may be more cost-effective.

## OUTDOOR SINGLE SLIDING PLATE SUSPENDED FROM AN E-W TRACK (ZERO ENERGY HOUSE SHUTTER) (Scheme 9.1)

This shutter was designed by S. Koch et al. for use in a solar house designed by T. V. Esbensen, V. Korsgaard, et al. The house is called Zero Energy House and is in Lyngby, Denmark. The shutter has performed excellently even in periods of extreme cold and high wind. For these reasons, I describe the shutter in detail despite its complexity and high cost.

### General Description

As shown in the following sketch, the house in question has three south windows, each 1.3 m × 1.3 m, and each equipped with a thick, insulating, external shutter. Each shutter is suspended from an overhead E-W track, and (by means of a hand-operated crank inside the room) can be moved east or west. Each, when in closed position, is fully edge-sealed.

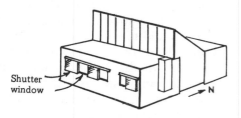

Shutter
window

Above each shutter there is a projecting wooden canopy, or housing.

Within the southernmost (outermost) part of the canopy there is a roll-up shade used in summer to reduce the amount of solar radiation reaching the window.

Each window proper can be opened by sliding it east or west along a horizontal rail just beneath the window. In winter, when the window is closed, it is drawn a few millimeters northward so that it presses against sealing strips.

The assembly of window, window frame, window track, shutter, shutter track, shutter seals, and roll-up curtain is so complicated that it cannot be made understandable merely by a set of conventional drawings. However, with the aid of many simplified drawings one can understand the design.

The accompanying perspective sketch of the shutter proper will probably help the reader very little. The top and bottom seals are only barely visible. The west end seal is entirely concealed. The east end seal (integral with the fixed housing, not with the shutter itself) is not shown.

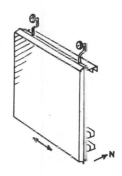

### Shutter Proper

The shutter proper is a square, flat assembly 1.3 m × 1.3 m × 5 cm. It consists mainly of a 5-cm-thick slab of polyurethane foam. The edges of the slab are protected by simple wooden framing members. The combination of slab and frame is contained within a pair of mating, rectangular, dished covers made of 2-mm-thick plastic (ABS plastic). The four projecting rims of one cover are joined to those of the other cover; these joined rims constitute four projecting fins which have no interesting function

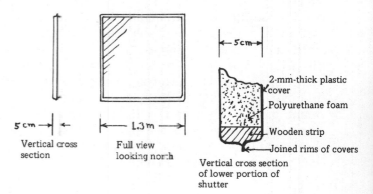

5 cm

Vertical cross section

— 1.3 m —

Full view looking north

— 5 cm —

2-mm-thick plastic cover

Polyurethane foam

Wooden strip

Joined rims of covers

Vertical cross section of lower portion of shutter

—for example, they play no part in guiding the shutter or in sealing the shutter to the window frame.

## Suspension

Attached to each end of the upper edge of the shutter there is a hanger. The two hangers bear the weight of the shutter and always hold the shutter at the same height above ground. Affixed to the top of each hanger is a ball bearing wheel (like a roller skate wheel) that can travel east or west along a horizontal channel, or track, that has a U-shaped cross-section (the "U" opens toward the north). The track is attached to an outer member of the fixed canopy.

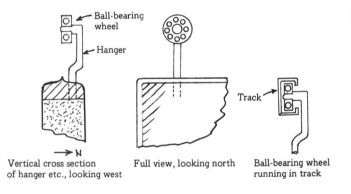

Vertical cross section of hanger etc., looking west

Full view, looking north

Ball-bearing wheel running in track

## Guide System for Lower Edge of Shutter

This system prevents motion of the lower edge of the shutter toward the north or south, while allowing the shutter to slide freely east or west. The system includes (1) a housed pair of 1-m-long, thin, horizontal, compressible strips (brushes) affixed, indirectly, to the shutter, and (2) a 2-m-long vane (affixed to the lower portion of the frame of the window) that is situated between the above-mentioned brushes and is lightly squeezed by them. The housing for the pair of brushes is a horizontal channel, the "U" of which opens up-ward; the channel is situated 3 cm north of the shutter and is affixed to it by means of brackets. Each brush consists of a horizontal linear array of thousands of 15-mm-long horizontal nylon fibers,

each of which is parallel to the north-south direc-tion. The two brushes oppose each other, and the vane intervenes. When the shutter is moved west, say, the brushes move with it and the vane remains stationary.

Note: the brushes shown in the following sketches are drawn poorly. In fact, a brush resem-bles a slender strip cut from a deep-pile rug.

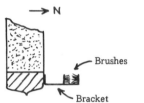
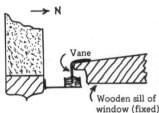

Vertical cross section of portion of shutter and brushes, looking west

Vertical cross section, looking west, of lower part of shutter, vane, and fixed sill of window

## Edge Seals

The edge seal for the upper edge of the shutter employs a brush, but no vane. The brush, which extends east-west, is situated 3 cm. north of the shutter and is affixed to it by means of brackets. The brush fibers aim downward and bear against the upper surface of the upper member of the movable window frame. Note that the brush con-tacts the frame in such a way as to offer practically no resistance to slight northward or southward motion of the window frame (as when the window clamp is being tightened or released).

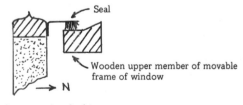

Vertical cross section, looking west

The edge seal for the lower edge of the shutter is a mirror image of the upper edge seal. The bracket is mounted a few centimeters above the lower edge of the shutter so as to allow room for the guide system discussed in a previous paragraph.

The edge seal for the east edge of the shutter consists of a long, thin, vertical brush that is attached to the east member of the fixed window frame, aims west, and lies in the plane of the shutter.

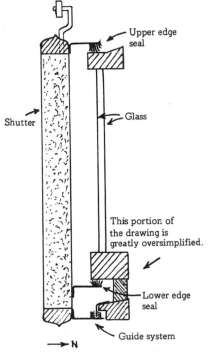

Shutter

Upper edge seal

Glass

This portion of the drawing is greatly oversimplified.

Lower edge seal

Guide system

N

Vertical cross section of shutter and window, looking west. Not to scale.

The edge seal for the west edge of the shutter consists of a long, thin, vertical brush that is situated 3 cm north of the shutter and is affixed to the shutter via brackets. The operation of closing the shutter presses this brush against the west face of the west member of the movable frame of the window.

The diagram at the bottom of the page shows the relationship between the east and west seals. Notice that they lie in different vertical planes. Notice that the west brush moves with the shutter (the upper and lower seals do so also), but the east brush does not.

In summary: When the shutter is closed, all four edges are sealed (despite the fact that one of the brushes lies in a different vertical plane from the other three). All are sealed by brushes. Two brushes (upper and lower) are always engaged, and their motions are sliding, tangential motions; the other two brushes are engaged only when the shutter is closed, and their engagement consists of a butting, or head-on pressing. Three of the brushes are attached to the shutter, hence are movable; the fourth is attached to the fixed frame of the window.

## Propulsion

The shutter is propelled westward (to open) or eastward (to close) by means of a manually operated crank situated in the room, near the lower

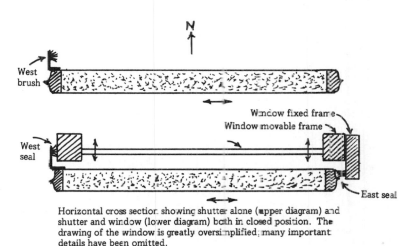

N

West brush

West seal

Window fixed frame

Window movable frame

East seal

Horizontal cross section showing shutter alone (upper diagram) and shutter and window (lower diagram) both in closed position. The drawing of the window is greatly oversimplified; many important details have been omitted.

west corner of the window. The crank turns a shaft running to the top of the window assembly, and, by means of a universal joint and a worm-and-gear device, turns a sprocket wheel situated in a space near the top of the shutter and within the canopy. About 1.2 m from this sprocket wheel there is a second such wheel. These two support and control a bicycle chain. One link of the chain is affixed to a bracket fastened (indirectly) to the upper edge of the shutter. A stopping device is provided to prevent the shutter from being driven too far westward. Proper tension in the chain is maintained by an idler sprocket.

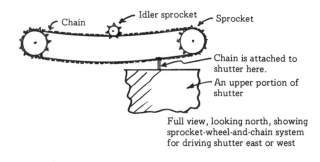

Full view, looking north, showing sprocket-wheel-and-chain system for driving shutter east or west

## Note Concerning Window Itself

The window is double-glazed. Its movable frame rests on two wheels, each of which is grooved to run on a single, slender, horizontal east-west rail (track). Although the wheels cannot move north or south (because the rail is fixed), the movable window frame is free to move a few millimeters north or south—by virtue of the fact that each wheel is free to slide north or south along its north-south axle. Thus after the window itself has been

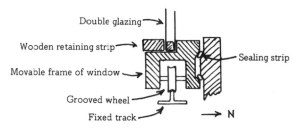

Vertical cross section, looking west, of lower edge of window. Before being clamped, the window (and its movable frame) is free to move one meter E and W and about 2 mm N and S.

slid shut, it can be pulled northward slightly so that it will press tightly against the sealing strips attached to the fixed frame. A special clamping system, manually operated from within the room, accomplishes this "northward tightening" of the closed window. Notice that although the window can be moved slightly north or south, the shutter cannot; yet all four seals continue to perform properly irrespective of the slight north-south motion of the window.

## Note Concerning Roll-Up Shade

Mounted beneath the south edge of the canopy there is a roll-up opaque shade which can be used in summer to exclude solar radiation—without the need for using the shutter (which, when closed, blocks all radiation, blocks all view, blocks passage of air).

## Performance

A typical double-glazed window of the Zero Energy House has the following heat transfer coefficient, according to *Solar Energy* 19, 195 (1977):

without shutter:

   3.10 W/(m² °C) i.e., 0.55 Btu/(ft² hr °F),

or R = 1.8

with shutter:

   0.40 W/(m² °C) i.e., 0.070 Btu/(ft² hr °F),

or R = 14.

## Discussion

Good features of the shutter:

   Very high R-value is achieved.

   Good edge seals are maintained—even if the shutter becomes slightly warped.

   The shutter is operated from inside the house.

   The shutter can never collect snow, never be highly stressed by wind.

   There is nothing that can fall.

Bad features:

Great complexity and great cost. I expect that these considerations alone would usually rule out use of these shutters.

Because the shutters cannot be closed without blocking sunlight, view, and passage of air, the designer must provide special additional devices for excluding solar radiation in summer (in this case, roll-up shades).

Operating the shutter (by turning a crank about 8 revolutions) takes some time.

Under special conditions, snow and ice might encumber the sealing strips and make them temporarily inoperative.

In summer, bees or wasps could nest behind the open shutters.

## OUTDOOR TANDEM SLIDING PLATES SUSPENDED FROM AN E-W TRACK (GOOSEBROOK HOUSE SHUTTER) (Scheme 9.2)

This shutter, invented by B. Anderson and C. Michal of Total Environmental Action, is in use at Goosebrook House in Harrisville, N.H. The shutter is huge and thick and provides high R-value; it is stored between cars in the garage.

The shutter includes two plates (east plate and west plate) in tandem, running on the same overhead east-west track (like a New England barn door track). Each plate is 7 ft high, 8 ft wide, and 3 in. thick. It includes a 1½-in. slab of Styrofoam sandwiched between 7/8-inch-thick, rough-sawn spruce boards. It weighs 400 lb. It is supported by two steel hangers equipped with ball bearing wheels which engage the overhead east-west track.

The shutter serves an all-glass Thermopane southwall area 7 ft high and 16 ft long. Good seals are provided. At the top a slender east-west strip of carpeting presses against the south face of the channel. At the east edge there is a neoprene weatherstrip. At the bottom a fin (part of angle iron) projecting downward travels in the space between two of the east-west decking boards. Besides providing a seal, this combination prevents the lower part of the shutter from swinging away from, or toward, the house.

When not in use, the shutter is far to the west, in the garage. The east-west track extends into the garage, in the space between the two cars; the two plates are parked in tandem there.

## Comment

The shutter reduces heat-loss enormously. Also it is durable and nearly vandal-proof. When parked, it is out of the way and out of the wind. It can be used in summer to exclude radiation and heat.

To open: The west plate is pushed to the left into the garage; the east plate likewise.

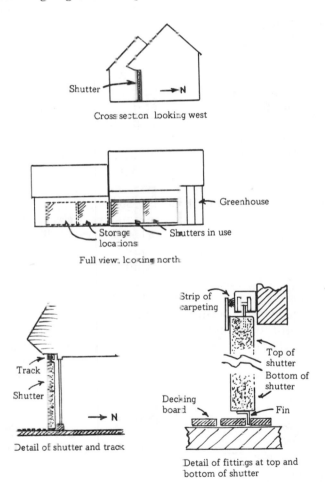

Cross section looking west

Full view, looking north

Detail of shutter and track

Detail of fittings at top and bottom of shutter

## Limitation

The very heavy plates are difficult to move. They could be made much lighter, I believe.

## OUTSIDE SINGLE SLIDING PLATE THAT HAS COMPLIANT SUSPENSION, SLIDES WITH LITTLE FRICTION, AND IS CLAMPED TIGHTLY AGAINST WINDOW FRAME (Scheme 9.3)

This sliding shutter, unlike most sliding shutters, has a compliant suspension system and slides in near-frictionless manner.

The insulating plate contains a 2-in. layer of rigid foam between two thin protective covers. The edges are stiffened by sturdy wooden strips.

The plate is suspended by means of two hangers that are equipped with small wheels which run along an overhead east-west track fastened to the house. Each hanger is long and thin and flexible and is attached to the plate in such a way that it has much north-south compliance (but no vertical or east-west compliance); thus the upper edge of the plate is free to move 1/2 in. north or south, and the lower edge has no north-south restraint at all. Accordingly, the north-south position of the track relative to the window frame is not critical, and when the shutter is being moved east or west its sealing strips make no contact with the house and the move is practically frictionless.

After being brought into position in front of the window, the shutter is manually clamped tight against the window frame so that the sealing strips (affixed to the north face of the plate) trap the air between shutter and window. Each clamp, employing a helical steel tension spring with a knob, engages a slotted tab integral with the plate. Because the peripheral frame of the shutter is very stiff, two clamps suffice—one at the middle of the east edge and the other at middle of west edge. When the clamps are not in use, they hang in recesses in the window frame.

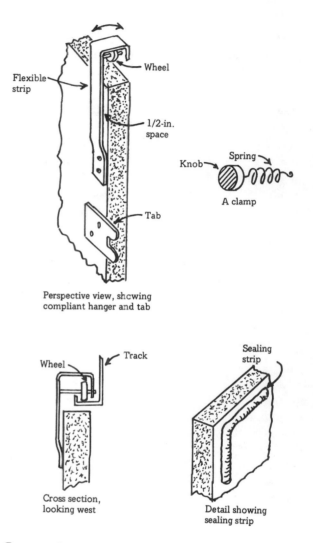

Perspective view, showing compliant hanger and tab

Cross section, looking west

Detail showing sealing strip

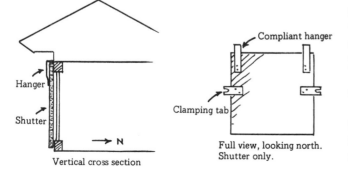

Hanger

Shutter

→ N

Vertical cross section

Compliant hanger

Clamping tab

Full view, looking north. Shutter only.

## Comment

Although the shutter is very easy to slide along, it can be sealed very tightly. Because the shutter has north-south compliance, no harm is done if the track is positioned imperfectly, or if the sealing strips are of different thickness than had been expected.

In summer the shutter can be used to exclude radiation and heat.

Also in summer the shutter can be tilted slightly (lower edge pulled 1 ft, say, southward) so as to act somewhat like an awning—to exclude radiation while admitting air and some light.

Note that the shutter leaves nothing exposed to collect snow and ice, and there is nothing to fall and hurt someone.

There is no frictional sliding (no rubbing-while-sliding); accordingly, prolonged use produces no wear.

To open: The clamps are released and the shutter is slid laterally into a "parking area" where restraining guide-arms keep it from swinging in the wind.

*Scheme 9.3a*

As above, except use four clamps. This eases the requirement on stiffness of the shutter.

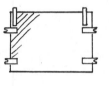

*Scheme 9.3b*

Revise the design of the track and the upper part of the hanger so that the shutter can swing through a 45-degree angle, to serve as an excellent awning in summer. (Swing the shutter down again before attempting to move it along the track.)

The shutter is held at 45 degrees by means of two 2-ft steel tie-rods, one at each end of shutter, which clip into place.

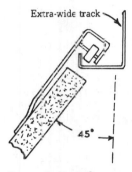

Shutter tilted at 45° to serve as awning in summer

# CHAPTER 10

# OUTDOOR OPAQUE SHUTTERS
# OF OTHER TYPES

- Outdoor Multi-Wooden-Strip Roll-Up Shutter That Has Swing-Out Capability (Rolladen Shutter Employing Wood) (Scheme 10.1)

- Outdoor Multi-Chambered-Hollow-Slat Roll-Up Shutter (Rolladen Shutter Employing Plastic) (Scheme 10.2)

- Outdoor Multi-Chambered-Hollow-Slat Roll-Up Shutter (Therma-Roll Shutter) (Scheme 10.3)

- Outdoor Multi-Chambered-Hollow-Slat Roll-Up Shutter (Serrande Shutter) (Scheme 10.4)

- Outdoor Multi-Chambered-Hollow-Slat Roll-Up Shutter (Pease Rolling Shutter) (Scheme 10.5)

- Outdoor Multi-Chambered-Hollow-Slat Roll-Up Shutter (Rolsekur Shutter) (Scheme 10.6)

- Outdoor Multi-Styrofoam-Board Roll-Up Shutter (Beale Shutter) (Scheme 10.7)

- Outdoor Venetian Blind Type Devices

Here I deal with shutters that are mounted outdoors, employ parallel sets of strips, slats, or bars, and can be rolled up or operated like a venetian blind.

Many of the designs employ almost the same design principles.

For indoor roll-up shutters, see Chapter 17.

## OUTDOOR MULTI-WOODEN-STRIP ROLL-UP SHUTTER THAT HAS SWING—OUT CAPABILITY (ROLLADEN SHUTTER EMPLOYING WOOD) (Scheme 10.1)

This shutter has been in widespread use throughout Europe for many decades.

The heart of the shutter is a set of wooden strips joined by means of two long, vertical, flexible tapes. The upper and lower edges of each strip are convex and concave (almost like tongue and groove), with the consequence that when the tapes are relaxed and each strip presses (by gravity) on its neighbor below, the strips make contact with one another and almost no light or rain can pass between them. From one strip to the next there is enough extra length of tape that, when the tape is made taut (by being pulled from above), small spaces exist between the strips, and much light and air can pass through.

The shutter proper (that is, the combination of strips and tapes) can be rolled up on a cylindrical roller that is mounted outdoors near the top of the window. The roller is turned by means of manually controlled cords that pass through the wall and into the room.

While the shutter proper is being rolled up or down, the ends of the strips slide vertically within rigid vertical channels (tracks).

The combination of shutter proper and channels is hinged at the top and can be swung outward and upward when and if certain restraining devices, operated from within the room, are released. Thus the device can become an awning. Such an awning is helpful on sunny days in summer: it excludes direct solar radiation, yet admits much diffuse light and affords view.

### Comment

Besides moderately reducing heat-loss in winter, the shutter prevents breakage of glass by vandals and discourages entry by burglars.

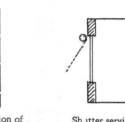

Cross section of          Shutter serving
strips when they          as awning
are close together

### Limitations

The reduction in heat-loss is modest.

The cost of the device and installation is high.

Bees or wasps might build nests behind the roller.

## OUTDOOR MULTI-CHAMBERED-HOLLOW-SLAT ROLL-UP SHUTTER (ROLLADEN SHUTTER EMPLOYING PLASTIC) (Scheme 10.2)

This shutter, made by American German Industries, Inc., of Scottsdale, Arizona, is much like the Rolladen shutter described above except as regards

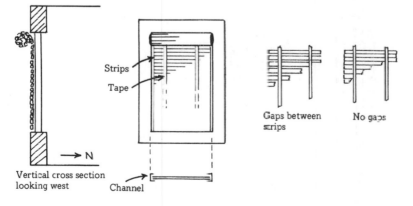

Strips
Tape

N

Vertical cross section
looking west        Channel

Gaps between      No gaps
strips

the slats. The slats are made of plastic (PVC) instead of wood. The cross section of the slat includes four closed chambers and one open chamber. The accompanying diagram shows the cross section. When the shutter is raised, i.e., rolled up, it forms a roll about 5 in. in diameter.

Cross section of
two joined slats

## OUTDOOR MULTI-CHAMBERED-HOLLOW-SLAT ROLL-UP SHUTTER (THERMA-ROLL SHUTTER) (Scheme 10.3)

This Therma-Roll Corp. shutter is somewhat similar to the Rolladen shutter. There are several varieties of Therma-Roll shutters; described here is one that is meant for retrofitting to existing buildings and is operated by a nylon tape rather than by a crank.

The shutter employs many long, thin slats. Each slat has a cross-section of about 1½ in. × 3/16 in. Each slat has four partitions, or walls, that define five long, thin, air-filled chambers. The first and last chambers include hooks for joining one slat to the next. The slat is of tough plastic (PVC) or foam-filled aluminum. The material is so tough that, with one's bare hands, one is virtually unable to damage a short piece of slat.

Where one slat is joined to the next there is enough play so that the set of slats can be rolled up on a 3-inch-diameter roller. Near the upper edge of each slat there are tiny slots. When the set of slats is in tension, the slots are exposed; they allow a very small amount of light and air to pass through. But when the set is in compression, as when lowered until the lowest slat is pressed downward onto the window sill by the weight of the upper slats, the slots are covered and little air can pass through.

The roller is mounted within an outdoor housing that is made of aluminum and has a square (5½ in. × 5½ in.) cross section.

The ends of the slats slide in vertical channels (neoprene seals within extruded aluminum channels) attached to the sides of the fixed frame of the window.

At the west end of the roller there is a pulley that guides a nylon tape (strap) 1/2 in. × 1/32 in. in cross section. The tape runs north into the room (via a 5/8-inch-diameter hole in the wall), then downward, and engages a wind-up roller situated just below the window sill.

Cost: About $7/ft$^2$ F.O.B. or, including installation, $9 or $10/ft$^2$.

*Scheme 10.3a*

As above, but the shutter is operated by means of a hand-operated crank. This option is available from the manufacturer.

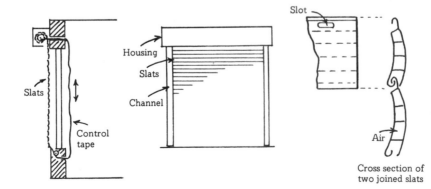

Cross section of
two joined slats

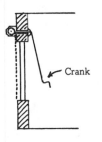

*Scheme 10.3b*

As above, except an electric motor is used to operate the shutter. This option also is available from the manufacturer.

*Scheme 10.3c*

In newly constructed buildings, instead of having the roller attached outside the house, the roller could be incorporated within the wall. A control tape or control crank may be used.

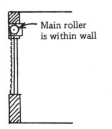

## OUTDOOR MULTI-CHAMBERED-HOLLOW-SLAT ROLL-UP SHUTTER (SERRANDE SHUTTER) (Scheme 10.4)

This shutter is made in Italy. In the United States it is distributed by various companies including Serrande of Italy, PO Box 1034, West Sacramento, CA 95691.

The design roughly parallels that of the Therma-Roll shutter. However, the individual PVC slat of the Serrande shutter is wider and thicker, as indicated in the accompanying sketch. The great thickness of the slat helps reduce heat-loss. Also it makes the slats sufficiently rigid that, even if the slats are as long as 6 ft., they require no special support or stiffening. When special stiffening is employed, lengths up to 10 ft are feasible.

The shutter may be raised or lowered manually by means of an indoor pulley system. Various arrangements of cranks, pulleys, and electric motors are available. There is a wide choice of color of slats. The shutter may be locked shut for increased protection against burglars.

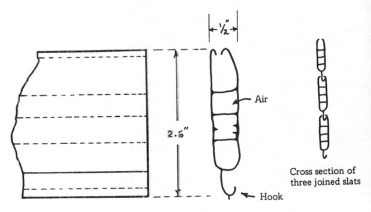

Cross section of three joined slats

## OUTDOOR MULTI-CHAMBERED-HOLLOW-SLAT ROLL-UP SHUTTER (PEASE ROLLING SHUTTER) (Scheme 10.5)

This outdoor roll-up shutter is made by Pease Co. of Indiana, a division of Pease Co. located in Ohio. There are two models of shutters: V and VM.

The slats of the Model V shutter are hollow and the 2⅛ in. × 1/2 in. cross section includes several chambers. The material is 0.040-in. PVC. Projections at the ends of the slats slide up and down in vertical channels that contain pile strip seals.

The shutter may be rolled up (on a steel roller) by means of an indoor strap and pulley system, which can be operated manually, with a crank, or by an electric motor. The rolled shutter is concealed within a housing of sheet metal or wood. Alternatively it may be situated within the wall, requiring no housing.

The Model VM shutter employs PVC slats that have a smaller cross section (1½ in. × 1/3 in.) and a thinner wall (0.030 in.).

The cost for a typical window is about $200 to $300 F.O.B. not including installation. The cost for a patio door 6 ft wide and 6 ft 8 in. high is about $550 F.O.B.

## OUTDOOR MULTI-CHAMBERED HOLLOW-SLAT ROLL-UP SHUTTER (ROLSEKUR SHUTTER) (Scheme 10.6)

This shutter has much resemblance to shutters described in the previous sections.

A typical version of the Rolsekur Corp. shutter employs hollow partitioned PVC slats which are joined by hook-like edges. Tiny slots near the upper edge of each slot are open or covered depending on whether the set of slats is in tension or compression. The vertical channels are of plastic or aluminum. Various locations of roller and housing are possible. Ordinarily the shutter is operated from indoors by means of a hand crank.

Some versions of the shutter have swing-out (awning) capability.

One version employs wooden slats.

Cost of typical type: $9 to $13/ft² F.O.B., not including installation.

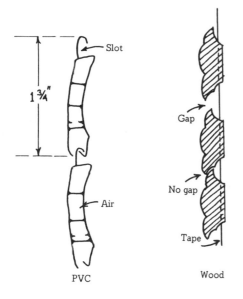

PVC                    Wood

## OUTDOOR MULTI-STYROFOAM-BOARD ROLL-UP SHUTTER (BEALE SHUTTER) (Scheme 10.7)

This shutter was invented, I understand, by Wm. T. Beale of Sunpower, Inc., of Athens, Ohio.

It makes use of a series of parallel coplanar boards of Styrofoam. Each board is a few feet long,

a few inches wide, and about 1 inch thick. It is connected to the next board by a strip of cloth. The assembly can be rolled up on a roller situated above the window, outdoors, under the eaves. The roller is operated from indoors by means of a cord running over a pulley.

When the shutter is unrolled, the edges (i.e., the ends of the boards) slide downward within closely fitting vertical channels. These prevent flutter and discourage flow of air into, or from, the space between the shutter and the glazing.

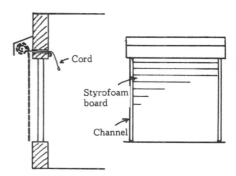

*Scheme 10.7a*

As above, but instead of rolling up the set of boards, they are stacked in a horizontal array just south of the top of the roller. (Based on information received from W. T. Beale.)

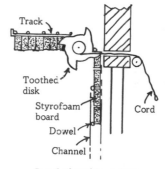

Detail of stacking system

## OUTDOOR VENETIAN BLIND TYPE DEVICES

Obviously, any venetian blind that is mounted in a snug manner just outside a window will somewhat reduce the heat-loss on winter nights.

But there are major difficulties:

Gaps may exist between the slats even when they are closed as tightly as feasible. Outdoor air can circulate into the space between blind and window.

Rain, snow, leaves, and insects may enter this space.

The device may swing and flutter in a strong wind.

The main appeal of such a device is that it can be put to very good use in summer. When closed, the blind excludes solar radiation. When partly open, it excludes most of the solar radiation but admits some daylight, allows some view, and provides ventilation.

Solar Science Industries sells an outdoor venetian blind, called Kool-Shade, used primarily to exclude solar radiation in summer. There are two types: standard, which has a relatively wide spacing of the vanes, and low-angle-sun, which has a closer spacing.

# CHAPTER 11

# BETWEEN-GLAZING-SHEETS DEVICES

- Avoiding Build-Up of Trapped Moisture
- Interposed Fixed Light-Transmitting Sheet (Scheme 11.1)
- Interposed Roll-Up Aluminized Sheet with Roller at Top (Scheme 11.2)
- Interposed Roll-up Aluminized Sheet with Roller at Bottom: Scheme Due in Large Part to N. B. Saunders (Scheme 11.3)
- Interposed Flexible Shade That Collapses Downward (Scheme 11.4)
- Interposed Flexible Shade That Collapses Upward (Scheme 11.5)
- Interposed Flexible Vertically Sliding Mattress (SUNLOC) (Scheme 11.6)
- Interposed Flexible Set of Rigid Boards (Insulider) (Scheme 11.7)
- Interposed Rigid Sliding Single Plate (Saskatchewan Shutter) (Scheme 11.8)
- Interposed Venetian Blind of Simple Type (Rolscreen Pella Slimshade) (Scheme 11.9)
- Interposed Quantity of Polystyrene-Foam Beads (Beadwall) (Scheme 11.10)
- Interposed Quantity of Hollow Glass Beads (3M Co. Macrosphere Window) (Scheme 11.11)
- Interposed Quantity of Wet Foam (Scheme 11.12)

The pros and cons of installing a thermal shutter or shade in the space between the inner and outer glazings are listed in Chapter 6. The main advantages are that such a device takes up little or no valuable space and is well protected. However, it is inaccessible for repair and moisture accumulation may be a problem unless suitable precautions are taken.

The moisture problem is discussed below. Then specific designs of insulating devices are described.

Note: Various devices discussed in other chapters—shutters and shades intended mainly for use indoors rather than between glazings—are pertinent here also. That is, some of them could be easily modified to work well in between-glazing-sheets locations. See, for example, various schemes described in Chapters 17–21.

## AVOIDING BUILD-UP OF TRAPPED MOISTURE

### The Problem

In winter, warm indoor air often contains more $H_2O$ than can be retained by cold air. Consequently when indoor air comes in contact with a cold pane of glass, $H_2O$ may be deposited, as a liquid, on this pane. More and more may be deposited. On a very cold night this water may freeze. When the outdoor temperature rises greatly, the ice may melt and water may run down onto the window sill and even onto the floor.

If, during a sunny daytime, the space adjacent to such a pane is opened to the room, such moisture may quickly evaporate.

But if a second sheet of glazing (or inside shutter) is installed and is moderately tightly sealed, trouble will arise: the air in the enclosed space may become especially cold, and moisture that accumulates there may stay there for days or weeks. Very little of it can find its way out.

The problem is especially acute if the indoor humidity is kept high, as by use of a humidifier, or by an in-house greenhouse that is watered frequently.

Build-up of moisture may be especially pronounced on windows on the leeward (downwind) sides of houses. Here the direction of air-leakage is outward: warm room air (containing much $H_2O$) finds its way into the cool space between glazing sheets; small pools of water or heavy frost may form. Windows at the upwind side of a house may suffer less build-up since the direction of air-leakage is from outdoors to indoors.

Why is it undesirable to have water accumulate between the shutter and window? Because:

The water may cause damage to window frames, sills, etc. Molds may grow there. (Aluminum frames, or frames of treated wood, may be unaffected.)

Water may drip onto the floor, producing stains and perhaps serious damage.

The accumulated moisture may freeze on a very cold night and the expansion that accompanies freezing may do physical damage.

If accumulated water encounters and wets porous insulating materials, their thermal conductivity will increase greatly.

The moisture may tend to evaporate and disappear on a hot sunny day if there are adequate escape paths for the moisture. But, alas, if there are such paths, much condensation may occur again on the next cold night.

### Some Solutions

Maintain low humidity in the house. If a humidifier is used, use it sparingly.

Make an extremely tight seal between the two sheets so that virtually no moisture can enter.

Open up the trapped air space on sunny days. Vent it to the room.

Connect the trapped air space with the outdoors via one or more tiny holes (weep holes) or cracks. Make the holes just large enough to allow moisture to escape at a fast rate on sunny days, yet not so large as to cause significant loss of heat by convection.

If moisture migrates through the body of the shutter or shade, paint it with a paint that prevents such migration (e.g., Insul-Aid paint).

Install, at the bottom of the trapped-air space, a horizontal strip of 3/4 in. × 1/2 in. foam or equivalent material that will soak up the condensate that runs down the glass. During the day, when the shutter or shade is removed and sunlight pours in, the moisture will evaporate from the foam. (I am indebted to Prof. Fuller Moore for telling me about his scheme.)

Waterproof the sill and install a special groove, gutter, or pipe that will deliver the condensate to a small bucket or to the outdoors.

Employ a desiccant. For example, use the scheme developed by R. K. Day and described in the following paragraph.

### Use of Small Balloon to Solve Moisture Problem of Sealed Double-Glazed Window

The problem of preventing accumulation of moisture in the space between so-called sealed double-glazed windows has been explored—and to some degree solved—by R. K. Day of Maumee, Ohio.

Moisture can enter such a space if (1) the seal between the two sheets of glass is imperfect, as is often the case, and (2) if large changes in temperature of the between-glazings air occur or if there are large changes in ambient atmospheric pressure. Air that enters the space in question may carry moisture with it, and any small supply of desiccant that may have been installed there earlier may, after a few years, become saturated and ineffective.

Day's solution (covered by U.S. patents 3,932,971 and 4,065,894 and by patents pending) is to install, outside the sealed space (but near it) a small rubber bag or balloon containing some desiccant that is connected to the between-glazings space by a slender tube. When changes in temperature or pressure cause air to flow from this space, the air flows into the balloon, expanding it slightly. Later, when conditions are reversed, air returns from the balloon to the between-glazings space. Thus it very seldom occurs that ordinary ambient air enters this space and accordingly moisture seldom enters. No appreciable build-up of moisture occurs, and any moisture that does enter the space is soon absorbed by the desiccant in the balloon.

The balloon may be concealed in a hollow within a member of the window frame—within an oversize bottom member, say. Or it may be concealed in a space adjacent to a side member or the top member. Of course, the balloon must communicate with the between-glazings space, e.g., via a slender tube.

If the balloon is concealed within a frame member, no daylight strikes it and it experiences no very high temperature. Therefore its lifetime may be long, e.g. a few decades.

Note: If the desiccant in the balloon ever becomes saturated, the balloon can be removed and filled with fresh desiccant. Or a new balloon (with fresh desiccant) can be installed. If there is already moisture between the two sheets of glazing, a newly installed balloon-with-desiccant will soon remove the moisture.

### INTERPOSED FIXED LIGHT-TRANSMITTING SHEET (Scheme 11.1)

An extremely cheap and easy way of cutting heat-loss to a moderate extent is to install a sheet of thin, transparent plastic between the outer and inner glazing. In this location the sheet is automatically protected from mechanical damage, UV radiation, and dust. Because of its protected location, it does not present a fire hazard. Because it is not exposed to outdoor winds or to indoor air currents, its edges do not have to be sealed; 1/4-in. gaps at the edges do little harm. Because it is transparent, it can be left in place night and day, all winter.

Such a sheet may reduce heat-loss by 15 to 25% (guess).

The sheet might cost only $1 and might be installed in 5 minutes.

How is the sheet to be attached? Almost any of the schemes described in Chapter 11 (for an outdoor added sheet) is applicable. Specifically, the sheet may be:

taped directly to the glass pane so that it lies close against it,

taped directly to the glass pane but held (by spacers) 1/4 in. from it,

taped directly to the frame of the sash,

secured to the top of the sash frame by means of a batten,

taped to a horizontal rod which in turn is secured to top of the sash.

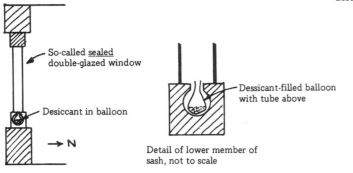

So-called sealed double-glazed window

Desiccant in balloon

→ N

Dessicant-filled balloon with tube above

Detail of lower member of sash, not to scale

attached to a frame which in turn is secured to the sash, or

attached to the jambs, i.e., at the sides, with the result that either sash may be raised or lowered without moving the added sheet.

Notice that, for most of these methods, the sheet is attached at the top only. The lower part of the sheet hangs free, i.e., dangles.

If the lower part tends to curl, the lower edge may be taped to a slender horizontal rod or strip of wood or piece of 1/8-inch-diameter steel wire. Such a device helps weigh down the lower part of the sheet and keep it flat.

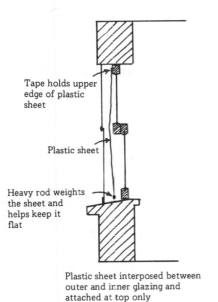

Tape holds upper edge of plastic sheet

Plastic sheet

Heavy rod weights the sheet and helps keep it flat

Plastic sheet interposed between outer and inner glazing and attached at top only

### Scheme 11.1a

As above, except a sheet of Foylon or other material that strongly reflects far-IR radiation is used. Such materials are listed in Appendix 7. They are somewhat expensive (about $1/ft$^2$), but may save about twice as much heat as ordinary plastic sheets save.

### Scheme 11.1b

As above, except at the other extreme. Use a sheet of plastic of the very cheapest kind, a sheet of thin, translucent plastic. This cuts the cost of the installation slightly, but view is sacrificed. In little-used rooms, view is unimportant.

### Scheme 11.1c

As above, except two sheets of plastic are used, spaced about 1/4 in. apart. To install two sheets may actually be easier than to install just one, as is made clear below; yet the saving of heat is almost doubled.

The procedure is to drape an extra-long sheet of plastic over a long horizontal rod and secure the rod to the upper part of the sash or to the jambs.

I learned of this general approach from R. K. Day, of Maumee, Ohio, who reports that he has tried out such schemes and found them to work well.

A possible improvement is to arrange the plastic sheet in the form of a continuous loop (by taping the overlapping ends together) and provide a heavy bar that rests on the bottom of the loop. The bar insures that the sheets are flat and vertical.

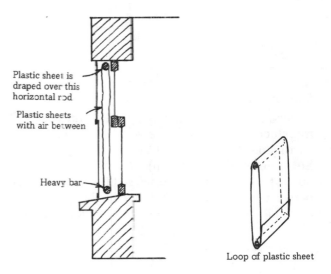

Plastic sheet is draped over this horizontal rod

Plastic sheets with air between

Heavy bar

Loop of plastic sheet

### Scheme 11.1d

If there is difficulty in making the above-described loop of plastic sheet fit nicely laterally, i.e., so that it extends very close to each jamb, one may use this procedure: prepare two loops, each about 60% as wide as the window. Employ a fixed upper bar and a suspended lower bar, as described above. Slide one plastic loop west until it rests snugly against the west jamb and slide the other against the east jamb. Thus the fit is snug at both sides, yet no care was needed in measuring the width of the window or cutting the plastic sheets to size. It may be desirable to cut the weighting bar into two segments, each serving just one loop. If the upper horizontal rod sags slightly at the center, this may help insure that the loops fit especially snugly against the jambs.

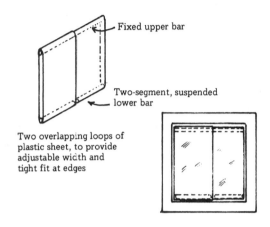

Two overlapping loops of plastic sheet, to provide adjustable width and tight fit at edges

### Scheme 11.1e

Instead of using very flexible plastic films, use stiff translucent sheets of low-cost type, such as flat or corrugated fiberglass-and-polyester (e.g., Kalwall Sun-Lite or Vistron Filon). Use two sheets, each 90% as wide as the window. After inserting them between the regular window and the storm window, move one west against the west jamb and the other east against the east jamb. Arrange for the upper edges of the sheets to be in contact with each other and the storm window, and arrange for the lower edges to be 1 in. apart. Thus several regions of trapped air are defined.

Fairly high R-value may be achieved. The sheets admit solar radiation and may be left in place night and day. They can be removed at any time in a few seconds. Installation is simple because there are no close tolerances and the sheets are simply laid in place, resting on the sill. It suffices just to open the lower sash—the sheets can be slid into the confined space by bending them slightly.

The main drawback is that the sheets are translucent, not transparent. View is lost.

### Scheme 11.1f

As above, except have one (or both) of the sheets of transparent material. Then the sheets can be left in place night and day without interfering with view. However, the transparent sheets are somewhat more expensive.

### Scheme 11.1g

For economy, use translucent sheets; but to provide at least some view, cut a small hole in each sheet and arrange for the holes to line up. If the holes are only about 2 in. × 8 in., they scarcely affect the heat-saving. If they are much larger than this, it may pay to cover them with a taped-on piece of transparent plastic.

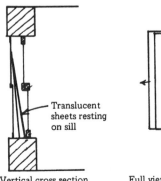
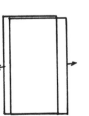

Vertical cross section looking west

Full view of the two overlapping translucent sheets

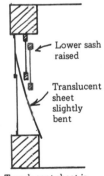

Translucent sheet is bent slightly during installation

## INTERPOSED ROLL-UP ALUMINIZED SHEET WITH ROLLER AT TOP (Scheme 11.2)

Here I describe a shade somewhat similar to one designed and used successfully by D. Kelbaugh of Princeton, N.J.

At the top of the space between the glazings there is a spring-return roller on which an aluminized sheet is mounted, e.g., a sheet of Foylon, Astrolon, or Dura-Shade. The roller has no pawl.

To close the shade, one pulls on a cord which causes the sheet to unroll and descend until the leading edge strikes the sill. The string, which runs under a pulley, is then secured by a jamb roller or other jamb device.

The interposed sheet is aluminized, with the aluminized face toward the room. It can be rolled up from a roller mounted at the bottom of the window. The roller contains a spring return but no latch (no pawl). To haul up the sheet, one pulls on a 1/32-inch-diameter nylon cord that runs through a curved passage (curved tube) in the top member of the fixed frame of the window. To keep the sheet up, the cord is secured in a jamb cleat situated close to the tube. The east and west edges of the sheet bear against slightly convex, vertical, filler strips of wood. At the top, good seal is provided by the sheet's leading-edge stiffening stick, which presses against a strip of foam rubber.

To open: Release cord from jamb cleat. The torque provided by the spring within the roller will cause the sheet to roll around the roller.

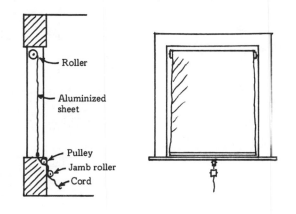

*Scheme 11.2a*

As above, except include channels at the sides to discourage air circulation.

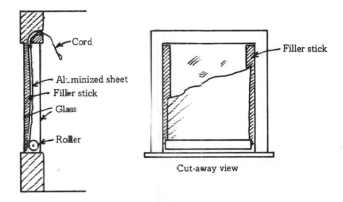

Cut-away view

Note: In these diagrams the jamb device is not shown.

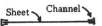

## INTERPOSED ROLL-UP ALUMINIZED SHEET WITH ROLLER AT BOTTOM: SCHEME DUE IN LARGE PART TO N. B. SAUNDERS (Scheme 11.3)

This scheme is based in large part on recent inventions by N. B. Saunders of Weston, Mass. Pertinent are (1) device displayed by him in Hartford, Conn., on 9/9/77, (2) his 1976 U.S. patent 3,952,947, (3) an additional patent pending, (4) various private communications.

### Limitations

The glass sheets must be at least 1 or 2 inches apart.

Access to the sheet and roller is difficult.

Most of the edge seals are only moderately tight.

Moisture may accumulate.

*Scheme 11.3a*

As above, but put the interposed aluminum sheet to additional use as part of a system for salvaging heat that otherwise might be lost. The system is applicable to greenhouses, schools, or other buildings in which large quantities of fresh air are required and exhaust fans are used to expel old air—causing the pressure in the building to be very slightly below atmospheric pressure.

Arrange for slight in-leak of outdoor air to occur just south of the interposed aluminized sheet. The air must pass downward in the airspace just south of that sheet, then upward in the airspace just north of it, and thence into the room. Heat that was likely to pass outward through the window structure is picked up and salvaged by this airstream.

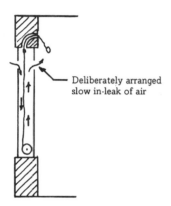

Deliberately arranged slow in-leak of air

A somewhat similar scheme was proposed some years ago by N. B. Saunders, I believe. Also, a related scheme was actually used in Solar Wind House at the University of Massachusetts, in Amherst. On. p. 26 of the June 1977 *Alternative Sources of Energy* (issue No. 26) one reads: "Fresh air is brought in through triple-glazed windows, fitted with a filter, which permit the warming of the incoming air by the heat normally lost through the window."

## INTERPOSED FLEXIBLE SHADE THAT COLLAPSES DOWNWARD (Scheme 11.4)

In 1974 I described, in an informally distributed report, a window insulating scheme employing, between the glazing layers, a vertical quilt which can be vertically collapsed merely by releasing a cord. The cord, which runs over a pulley near the top of the window, controls a heavy horizontal bar,

e.g., of steel, from which the quilt is suspended. When the bar is released, it descends, collapsing and compressing the quilt below it. When the bar comes to rest, it is situated only a few inches above the sill and the quilt lies crumpled and compressed just below the bar. The system is cheap and durable.

The cord is attached to an ear that projects above the center of the bar, and accordingly the bar tends to be level (horizontal) when suspended by the cord.

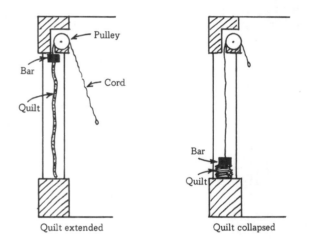

Quilt extended                    Quilt collapsed

*Scheme 11.4a*

As above, except use a set of four sheets of thin cloth or plastic instead of the quilt. The crumpled sheets may occupy less space than a crumpled quilt. One or more of the sheets might be faced with aluminum foil, to further reduce nighttime heat-loss by radiation. It is assumed that the four sheets will have enough inherent waviness, or enough fuzz, so that there will be airspaces between them.

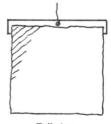

Full view

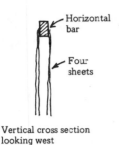

Horizontal bar

Four sheets

Vertical cross section looking west

They may provide greater heat saving than a quilt provides.

(In 1977 C. G. Wing of Cornerstones made a shade that could be collapsed vertically. However, it was situated indoors, rather than between the glazing sheets. It was found to perform well.)

### Scheme 11.4b

As above, except fasten (by means of a batten, say) the lower edge of the quilt (or set of sheets) permanently to the sill. Then no circulation of air beneath the lower edge can occur. This modification was suggested by C. G. Wing in about 1978.

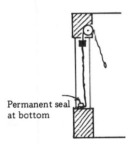

Permanent seal at bottom

### Scheme 11.4c

As above, except at the top, use a piece of pipe instead of a rectangular-cross-section bar, and drape an extra long quilt or long sheet of plastic over it to form two separate vertical barriers with trapped air between them. Glue the quilt to the pipe. This scheme reduces assembly time.

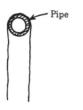

Pipe

### Scheme 11.4d

Instead of using a simple, flat quilt, use a quilt or sheet that has been preformed so as to fold in a systematic zig-zag manner. Such a device will collapse into a smaller space. Also when this shade is in use it will be especially effective because the zig-zag shape tends to discourage vertical circulation of air.

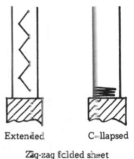

Extended    Collapsed

Zig-zag folded sheet

### Scheme 11.4e

Instead of using a quilt or set of nearly flat sheets, use a honeycomb structure that includes many layers and collapses vertically in accordion manner. This scheme was proposed to me in 1977 by George Sanders of Stockbridge, Mass.

Note: The use of a collapsible honeycomb structure situated in the room, rather than between glazing sheets, has been proposed by J. J. Anderson, Jr., of Ramsey, N.J. For details, see Chapter 19.

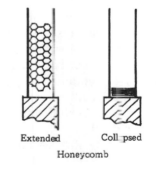

Extended    Collapsed

Honeycomb

### Scheme 11.4f

If the device is to be used in a double-sash window and if the space at mid-height is somewhat constricted, include—just below the mid-height centerline of the sheet—another horizontal bar. This will help insure that when the sheet is allowed to descend it will do so, i.e., it will not become crumpled and jammed in the constricted region. In summary, there are two massive horizontal bars: one to insure easy downward passage through the constriction and the other to compact the fully descended sheet.

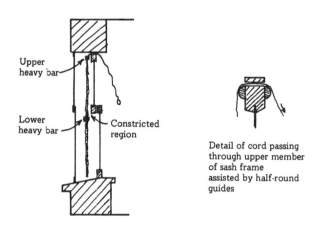

Detail of cord passing through upper member of sash frame assisted by half-round guides

## INTERPOSED FLEXIBLE SHADE THAT COLLAPSES UPWARD (Scheme 11.5)

This shade was proposed by D. S. Kelbaugh of Princeton, N.J. He had in mind using such a device to serve a Trombe wall. There are several sheets of flexible aluminized cloth arranged in series, and all are permanently attached to the upper part of the fixed frame of the window.

To open the shade, one hoists up the lower parts of the sheets by means of several cords. The set of cloths then forms a compact bundle near the top of the window. The cords are secured by a jamb roller or cleat. If a moderately heavy bar is incorporated in the lower edge of the set of cloths, a good seal at the bottom is achieved when the cloths are allowed to descend fully. If the cloths are slightly wider than the window, they will press against the jambs; thus seals at the sides will be provided.

### Limitations

When pulling on the cords, one must be careful to pull them together simultaneously.

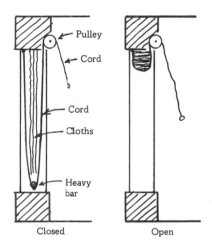

The raised assembly may sag considerably between the hoisting locations. Battens may be added to prevent this.

## INTERPOSED FLEXIBLE VERTICALLY-SLIDING MATTRESS (SUNLOC) (Scheme 11.6)

The best features of this shutter were invented by Leandre Poisson of Solar Survival, Harrisville, N.H. Many variations have been developed. The trademarked name *Sunloc* has been used.

The heart of the shutter is a 4-inch-thick flexible plate (mattress) of plastic foam. On winter nights this is situated between 4-inch-apart glazing sheets of Kalwall Sun-Lite or the equivalent. On sunny winter days the mattress is situated, folded, in a box adjacent to the base of the window. The mattress is moved from use position to parked position or vice versa by means of ropes: an opening rope near the top and a closing rope near the bottom. Large- diameter rollers within the box guide the mattress. Small pulleys guide the ropes.

### Comment

When the mattress is in position of use, it presses against both glazing sheets. There are two face seals. the overall R-value may be very high.

### Limitations

The installation must be planned before the wall is built.

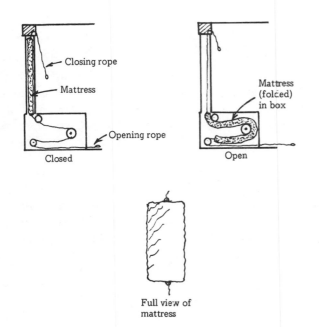

Closed — Open

Full view of mattress

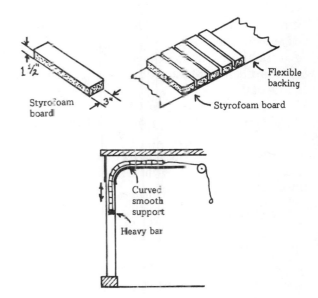

The box takes up much space (but can be used as a window seat or as a platform for plants).

*Scheme 11.6a*

*As above, except store the mattress in a box near the ceiling.*

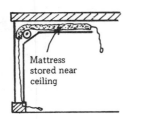

Mattress stored near ceiling

*Other modifications*

A wide range of mattress thicknesses may be used, and a wide range of glazing materials (e.g., Plexiglas, glass). The mattress may be parked below the floor or above the ceiling.

## INTERPOSED FLEXIBLE SET OF RIGID BOARDS (INSULIDER) (Scheme 11.7)

This device was developed by Jerry Jaksha of Lincoln, Nebraska. He calls it *Insulider*.

Many 1½-inch-thick, 3 in. wide boards of Styrofoam SM are glued to a tough, flexible backing sheet, e.g., rip-stop nylon cloth. A heavy horizontal bar is attached to the lower edge of the assembly to help it descend. A curved support system is installed close below the ceiling.

To open the shutter, one pulls on a rope that is attached to the upper edge of the flexible assembly and runs over a pulley, then secures the rope end by a cleat or hook.

It has been found helpful to extend the backing sheet over the ends of the boards, with cuts between them so that the flexibility of the assembly is not impaired. The cuts are made after the entire gluing operation is completed.

## INTERPOSED RIGID SLIDING SINGLE PLATE (SASKATCHEWAN SHUTTER) [Scheme 11.8]

The use of a vertical plate of rigid foam that may be slid to the right or left into a recess in the wall was demonstrated early in 1977 by the designers of the Saskatchewan Conservation House. (Sponsor: Saskatchewan Office of Energy Conservation. Built under supervision of Saskatchewan Housing Corp. Architect: Henry Grolle.)

In the 12-inch-thick south wall a 4 ft × 4 ft window was installed. Between the outer (south) glazing, which is of conventional type, and the inner (north) glazing, which is part of an Andersen casement, there is a 4-in. space. At night there is a 3-inch-thick shutter in this space; it consists of Styrofoam with sheet metal sheathing. The R-value of this device is 15.

In the wall region just to the west of the window there is a boxed-in 4-inch-thick recess, or slot, that can receive the shutter. On the south and north sides of the slot enough high-quality insulation is present to provide an R-value of 40, to match that of other regions of the wall.

To open the shutter, one opens the casement, grasps a handle recessed in the east end of the shutter, and pulls the shutter westward. Although well weatherstripped, the shutter slides easily; no rollers are provided. Small indentations on the north faces of the shutter facilitate pushing it with the fingers toward closed or open position. (An earlier plan, later dropped, was to move the shutter with the aid of a continuous draw-cord.)

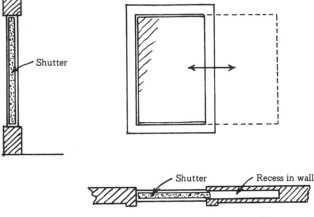

*Scheme 11.8a*

As above, except arrange to have the plate slide upward or downward into a recess above or below the window. Obviously, this scheme is feasible only if the window is small and there is much free space above or below the window. Preferably the plate should be very lightweight.

## INTERPOSED VENETIAN BLIND OF SIMPLE TYPE (ROLSCREEN PELLA SLIMSHADE) (Scheme 11.9)

This device consists of a double-glazed casement window called *Rolscreen Pella Energy-Tight Window* and a venetian blind called *Rolscreen Pella Slimshade*. Both are made by Rolscreen Co. of Pella, Iowa.

The casement-type window, with its two sheets of glass 13/16 in. apart mounted in a wood frame, is opened or closed by means of a small crank. The venetian blind is mounted in the between-glazing-sheets space, which is vented to outdoors to allow moisture to escape.

The 11/16-inch-wide slats, which are slightly curved in cross section, are of 0.010-in. aluminum coated with white or bronze-colored paint. Spaced 19/32 in. apart, the slats are supported by four vertical polyester cords about 1/16 in. in diameter. To rotate the slats, one turns a small knob in a lower corner of the window. A small hole has been drilled in the glass to accommodate the shaft of the knob. The slats cannot be raised or lowered.

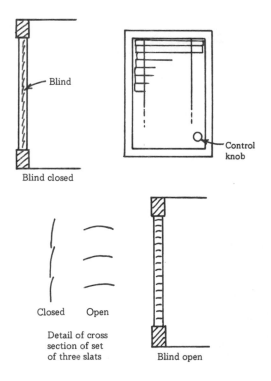

Closing the venetian blind is said to reduce the U-value of the window from 0.49 to 0.41. The corresponding R-values are 2.04 and 2.44.

In summer the blind may be used to exclude most of the incident solar radiation.

Cost: The incremental cost of such a blind for a window about 3 ft × 2 ft is about $20., i.e., about $3.50/ft².

## Comment

The special venetian blind can be installed in many modern types of double-hung windows and in French-type sliding glass doors. Retrofitting is feasible: local distributors are equipped to drill the necessary hole in the glazing and complete all phases of the installation. The Rolscreen Pella Energy-Tight window is designed so that one glass sheet can be removed readily to permit cleaning the glass and removing and cleaning the set of vanes.

*Scheme 11.9a*

As above, except use vanes such as were designed by T. E. Johnson et al. for use in MIT Solar Building V. The vanes are aluminized and are used during sunny winter days to reflect solar radiation toward thermal storage tiles integral with the ceiling. The vanes can be fairly tightly closed on winter nights to reduce heat loss and may be used also to exclude solar radiation on hot days in summer. (See Ref. S-235aa, p. 105.)

See also p. 180 dealing with venetian blinds for use indoors.)

## INTERPOSED QUANTITY OF POLYSTYRENE-FOAM BEADS (BEADWALL) (Scheme 11.10)

This device, applicable to houses, large buildings, greenhouses, etc., was invented in about 1972 by David Harrison (U.S. patent 3,903,665) and was later improved and marketed by Zomeworks Corp.

A Beadwall window has two large parallel glazing sheets which are spaced about 2½ to 4 in. apart. Each sheet may be 3 or 4 ft wide and 7 ft high

and may be of 3/16-in. tempered glass, or Kalwall Sun-Lite, or other material. At night the space between the sheets is filled with millions of tiny white beads of polystyrene foam, each about 1/16 in. to 1/8 in. in diameter. Early in the morning of a sunny day the beads are expelled from the space. They are transferred via a 1½-in. PVC pipe to a storage tank, e.g., a vertical cylindrical fiberglass tank of 17 to 28 ft³ capacity. The beads are transported by high-speed airstreams driven by two small blowers, one near the window outlet and the other near the tank outlet. The transfer takes 1/2 minute. The beads have been treated with flame retardant and also with an antistatic agent that keeps them from clinging to the glazing. Transfer is usually controlled manually, but may be automated. Filling or emptying the window may be halted at any stage, to leave part of the window clear for solar heating, view and illumination and part filled for insulation and privacy. Normally the window is vertical, but the system is applicable to systems tilted as far as 45° from the vertical.

In summer, the operation may be reversed: the window may be filled during the day to exclude solar radiation and ambient heat and, at other times, may be kept empty to let indoor heat escape.

R-value of window with 2½-in. space: 8 when full, 1.5 when empty.

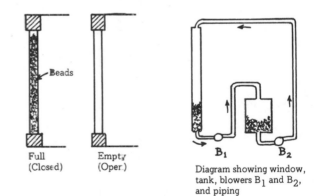

Full (Closed)    Empty (Oper.)

Diagram showing window, tank, blowers $B_1$ and $B_2$, and piping

## Limitations

The cost is high.

Will the inaccessible faces of the glazing sheets become dirty or cloudy?

If the anti-static tires, the beads may not fully drain.

The storage tank takes up much space.

If the exit hole from the window were to become clogged and air-pressure from the blower were to build up here, would the glazing burst?

Note: Some difficulties and special solutions are described by F. Kjelshus in Ref. I-402f, p. 750. Some revised designs have been developed by Egge Research of Kingston, New York, and by D. P. Greider of Solar Central in Mechanicsburg, Ohio.

## INTERPOSED QUANTITY OF HOLLOW GLASS BEADS (3M CO. MACROSPHERE WINDOW) (Scheme 11.11)

This window, announced in mid-1978 by 3M Company, is somewhat like Beadwall. It employs what is called "3M Co. Portable Insulation," said to consist of hollow beads of foam glass (75%) and phenolic resin binder (25%).

The R-value is said to be about 3 per inch.

Window partially filled with hollow beads

## INTERPOSED QUANTITY OF WET FOAM (Scheme 11.12)

Wet foam introduced between plastic glazing sheets can form excellent insulation, according to investigations made in 1975 by engineers at Arbman Development, AB, Kungsgatan 62, Stockholm, Sweden, as summarized in *Solar Energy Digest* of Feb. 1976. Using water, detergent, and a blower, foam is easily generated and driven into the between-glazings space. If appropriate additives are used, the foam is stable and long lasting. Inasmuch as the transmittance of the foam for solar radiation is fairly high (about 50%), it is sometimes feasible to leave the foam in place during the day. Clearing away the foam is a simple matter.

The system was developed mainly for use in greenhouses.

# CHAPTER 12

# INDOOR TRANSPARENT AND TRANSLUCENT DEVICES

- Indoor Single Fixed Sheet
- Indoor Device Employing Two Large Fixed Translucent Sheets: A Simple Sheet and a Sheet of Bubble Plastic (Scheme 12.1)
- Indoor Device Employing Two Sheets of Bubble Plastic and a Wooden Frame (Scheme 12.2)
- Indoor Device Employing Sheet of Corrugated Translucent Plastic (Scheme 12.3)

The variety of shutters and shades intended for indoor use is enormous.

This chapter deals solely with transparent and translucent devices. Although they are not (strictly speaking) shutters or shades, they perform somewhat the same function, and (like shutters and shades) are easily installed on existing windows.

They have the special merit that they can be left in place during the day as well as during the night, saving heat and money 24 hours a day and requiring no attention at all.

The general requirements are indicated in Chapter 6. Some of the more important materials are listed in Appendix 6.

Many of the methods of attachment listed in Chapter 6 are applicable here also.

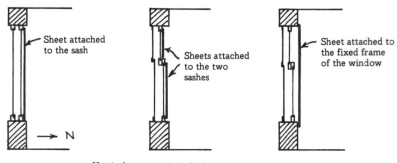

Vertical cross sections, looking west

## INDOOR SINGLE FIXED SHEET

### Glass

A sheet of glass that is to be permanently mounted indoors may be attached to the window sash or to the fixed frame of the window. The sheet should be mounted so as to leave a region of trapped air between it and the existing window glass. The optimum thickness of airspace is 1/2 to 1 in. A space of several inches is slightly inferior as it permits easier circulation of air in the airspace and thus encourages heat-loss by convection. A space of much less than 1/2 in. is far inferior.

The added sheet of glass may be attached by any of several means described in Chapter 6 dealing with added outdoor sheets. However, some of those schemes may make it difficult to open or clean the window or may create problems of moisture condensation.

Note: If the window is of double sash type and the glass is to be attached to the face of the sash, it may pay to install the glazing on the indoor side of the lower sash and on the outdoor side of the upper sash. Then the window can be opened in a normal

manner, with no interference by the added glazing. (Suggested by A. Wade, Ref. W-40, p. 234.)

Alternatively, the added sheet may be installed in a specially prepared (rabbeted) recess in the existing sash. Using the scheme developed by S. T. Coffin of Lincoln, Mass., one employs a hand-held, electrically driven rabbeting tool to make a shallow cut in the members of the sash frame, or in the muntins, to provide a recess into which the sheet of glass (or rigid plastic) may be placed. The sheet is secured by means of glazing points and putty, or the glazing compound Geocel sealant may be used. A coat of oil base paint will help exclude moisture. The added sheet and edge seals are so inconspicuous that, on casually inspecting the sash, a visitor may be unaware that an additional sheet has been installed. The added sheet does not interfere with opening the window.

A slightly different scheme has been developed by Newton-Waltham Glass Co. of Waltham, Mass. The glass is held 3/8 in. from the existing

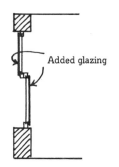

Added glazing

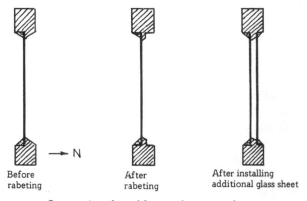

| Before rabeting | After rabeting | After installing additional glass sheet |

Cross section of wood-frame sash, not to scale

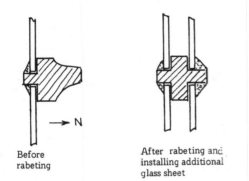

Before rabeting

After rabeting and installing additional glass sheet

pane by a slender mahogany frame which fits directly against the existing pane. The frame lies within the enclosure defined by the sash and is attached to the sash. A small amount of desiccant is included. Cost of materials: $3.25/ft². The company has filed a patent application.

## Rigid Transparent Plastic

Such a sheet may be of Plexiglas, Lucite, Lexan, or other materials. It may be mounted by any of the schemes described just above or described in Chapter 6. or one may drill holes through the edges of the sheet and attach it by means of screws.

Alternatively, one may purchase a rigid plastic sheet that is already mounted in a slender frame of plastic or metal. Such assemblies are available at many builders' supply stores can be installed easily by the homeowner. and can easily be removed when spring comes.

Dayton Corp. of Boston, Mass. (under license to Econ Inc.), sells a variety of add-on windows called Energlaze. Those intended for use on average size windows are of 1/8-in. Plexiglas. Snug-fitting frames, or channels, of ABS plastic are supplied they have a U-shaped cross section. Alternatively, framing members that have L-shaped cross sections may be used, and special clamping strips complete the mounting. The assemblies are held in place by thin magnetic strips or pressure-sensitive adhesive. (Note: If the sheet is attached magnetically it can be removed at any time relatively easily. If the sash is of wood, slender steel strips (to attract the magnetic strips)

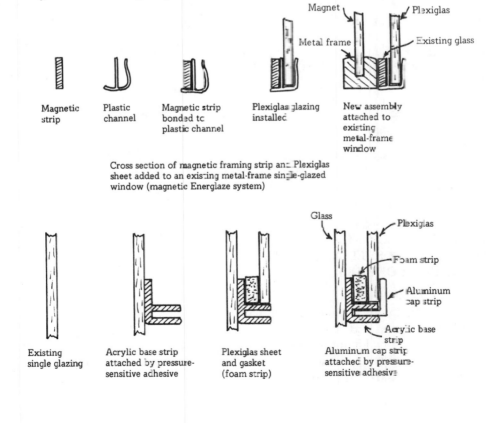

Magnetic strip

Plastic channel

Magnetic strip bonded to plastic channel

Plexiglas glazing installed

New assembly attached to existing metal-frame window

Magnet    Metal frame    Plexiglas    Existing glass

Cross section of magnetic framing strip and Plexiglas sheet added to an existing metal-frame single-glazed window (magnetic Energlaze system)

Existing single glazing

Acrylic base strip attached by pressure-sensitive adhesive

Plexiglas sheet and gasket (foam strip)

Aluminum cap strip attached by pressure-sensitive adhesive

Glass    Plexiglas    Foam strip    Aluminum cap strip    Acrylic base strip are added.) The typical cost, when 1000 ft² of such glazing is added to an existing building, is $5 to $7 per square foot.

Plaskolite, Inc., sells rigid plastic glazing sheets and snap-on vinyl framing strips. the system is called *In-sider Storm Window*.

## Flexible Transparent Plastic

Many kinds of flexible plastic sheets are listed in Appendix 6. Some are almost perfectly transparent. One kind that I have found to be inexpensive and almost perfectly transparent is:

Poly-Pane, a 0.001-in. sheet made by Warp Bros. of Chicago, Ill. It costs only a few cents per square foot.

Plaskolite, Inc., sells a "Weatherizer Kit" that includes a thin plastic sheet and suitable mounting strips.

Many ways of mounting such sheets on the outdoor side of a window are indicated in Chapter 6, and most of these schemes are suitable for indoor installation also. However, even simpler schemes may be appropriate for indoor installation inasmuch as there is no problem with wind or rain. The sheets may be attached with tape, which has the added virtue of providing an excellent seal. Alternatively battens may be used; even battens of cardboard may be effective.

## Flexible Translucent Plastic

Such material is even cheaper, but greatly impairs the room occupant's view of outdoors. Appendix 6 lists many kinds of translucent plastic sheets, and Chapter 6 describes many mounting methods.

## INDOOR DEVICE EMPLOYING TWO LARGE, FIXED, TRANSLUCENT SHEETS: A SIMPLE SHEET AND A SHEET OF BUBBLE PLASTIC (Scheme 12.1)

Two large, fixed, translucent sheets are employed: a sheet of bubble plastic and a simple sheet of tough plastic. The sheet of bubble plastic is attached by tape to a wooden frame which is mounted close to the existing glass. There is a

3/4-in. airspace between the bubble plastic and the glass. The simple sheet is then installed, by any of the means discussed in Chapter 11, in such a position that there is an airspace between it and the bubble plastic.

### Comment

The scheme is highly effective thermally because there are two layers of trapped air and also an array of tiny plastic bubbles. The system is durable because the sheet of bubble plastic is protected and the exposed sheet is of tough material. The wooden frame is not very visible, being partly obscured by the translucent tough sheet, and therefore can be crudely made. The two sheets can be left in place night and day throughout the winter. At night they provide privacy.

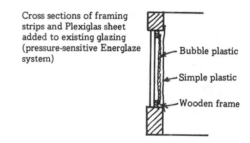

Cross sections of framing strips and Plexiglas sheet added to existing glazing (pressure-sensitive Energlaze system)

— Bubble plastic

— Simple plastic

— Wooden frame

### Limitations

View is obscured.

Moisture condensation may be a problem, but the presence of the bubble plastic may reduce or eliminate the problem.

*Scheme 12.1a*

As above, except omit the wooden frame. Tape the bubble plastic directly to the glass. This halves the

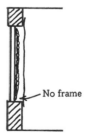

—No frame

cost and reduces the threat of moisture condensation, but the thermal performance is slightly inferior and removing the added sheets in the spring is slightly more difficult.

*Scheme 12.1b*

Omit the wooden frame. Tape some slender spacers (transparent drinking straws, or twisted strips of transparent plastic) to the south side of the simple plastic sheet and then attach the bubble plastic sheet to it, with the spacers serving to maintain a 1/4-in. airspace.

When and if moisture accumulates on the north glass sheet, merely detach one side of the pair of plastic sheets—on a sunny day—and the moisture will soon evaporate.

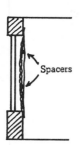

INDOOR DEVICE EMPLOYING TWO SHEETS OF BUBBLE PLASTIC AND A WOODEN FRAME (Scheme 12.2)

A 3/4-inch-thick wooden frame is prepared and a sheet of bubble plastic is wrapped around it—all the way around so as to provide two sheets of the material with a layer of trapped air between. The

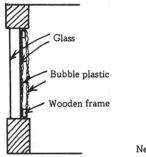

Near-full-scale cross section of part of wooden frame and wrapping of bubble plastic

sheet is secured with tacks, staples, or tape. The assembly is then installed in the window recess, i.e., between the jambs. It is pressed against the sash or directly against the glass and is held in place by friction or by means of clips or tape.

## INDOOR DEVICE EMPLOYING SHEET OF CORRUGATED TRANSLUCENT PLASTIC (Scheme 12.3)

A corrugated sheet of translucent plastic (for example, a sheet of Filon made by Vistron Corp.), oriented so that the corrugations are horizontal, is clipped or taped to the north side of the north glazing. At the center of the plastic sheet there is a hole (cut with a knife) 6-in. wide and 3-in. high, to serve as a clear-view peephole.

### Comment

Such a sheet is easy to install and can be left in place night and day. It can be removed in a few seconds and reinstalled almost as quickly. It is translucent and thus blocks view, but the small opening at the center provides a narrow "emergency" view. The reduction in heat-loss is only moderate but the cost effectiveness of the device is high. Because the sheet is applied directly to the north glass sheet, no tight seal is required.

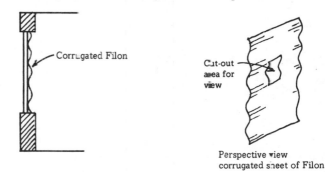

Perspective view
corrugated sheet of Filon

### Limitations

There is loss of broad view.

The corrugated sheet is flammable.

The corrugated sheet bends readily in one dimension and must be secured to the glass sheet at about six places.

*Scheme 12.3a*

As above, but install an additional sheet: a thin flat sheet of translucent plastic. Attach it to the room side of the corrugated sheet. This doubles the number of pockets of trapped air and nearly doubles the heat saving.

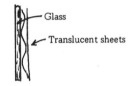

*Scheme 12.3b*

Use two corrugated sheets, back-to-back, with a thin sheet of translucent plastic between them. This creates even more pockets of trapped air. The corrugated sheets hold the thin, flexible (delicate) sheet in place and protect it from injury.

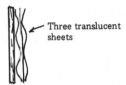

*Scheme 12.3c*

Use a plate of twinwall plastic instead of a corrugated sheet. The twinwall plate is stiffer and stays more nearly flat. Also it has higher transmittance —admits more solar radiation. It is more nearly transparent, but it costs more and is harder to cut to size.

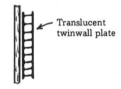

# CHAPTER 13

# INDOOR OPAQUE SHUTTERS THAT ARE NOT PERMANENTLY ATTACHED

- Special Benefits of Lightweight Shutters
- Great Appeal of a Thermax Plate
- Appeal of a Styrofoam Plate
- Plates of Other Materials
- Parking Problem
- Knobs and Handles
- Seals
- Indoor Simple Plate That Is Pressed Against Window and Rests on Sill (Scheme 13.1)
- Indoor Simple Plate Supported by the Edges of the Sash (Scheme 13.2)
- Indoor Simple Plate Supported by Beveled Plastic Strips Cemented to the Glass Itself (Scheme 13.3)
- Indoor Simple Plate Affixed to the Glass by Magnets (Scheme 13.4)
- Indoor Folding Shutter That Is Installed Symmetrically (Scheme 13.5)

Here we take up very-low-cost devices: indoor shutters that are not permanently attached.

They may cut heat-loss drastically, yet may be easy to make and install. True, removing them each morning and reinstalling them each evening is a chore. But the chore is completed in 10 or 20 seconds per window—and there is this bonus: during the day (and all summer) the shutter may be stored in some out-of-sight location, leaving the windows entirely unencumbered.

## SPECIAL BENEFITS OF LIGHTWEIGHT SHUTTERS

Many bonuses fall into the lap of a shutter designer who succeeds in keeping the weight of a shutter very low. Shutters that consist of little more than a plate of Thermax or Styrofoam SM are indeed very light and they provide these special benefits:

● If, while a not-permanently-attached shutter is being detached (at the start of a sunny day), it is accidentally released and falls some distance and lands on your foot, no injury is done and there is no pain. Likewise, if it lands on a playing child, or a sleeping dog, no harm is done. Nor is the shutter itself damaged.

● If, while a shutter that is hinged at the top is (when open) accidentally released and swings downward freely, no harm is done. the mass of the shutter is so small, and the wind resistance so great, that the shutter descends relatively slowly and delivers only a gentle blow.

● If the shutter is to be attached by hinges, cheap lightweight hinges suffice and they can be attached merely by means of sticky tape. if, some day, the hinges fail and the shutter falls, no harm is done.

● If the shutter is to be raised and lowered, no counterweight is needed. Even if the shutter is of large area, it can be raised easily, even by a child, without need for counterweight, pulley, rope. (If a counterweight is provided, it can be very light; thus it itself presents no hazard.)

● If spring clips are employed to hold the shutter open (against the force of gravity), simple weak clips suffice.

● If the shutter is of a horizontally sliding type, the sliding force required is small. Operation is almost effortless. No rollers are needed.

● Transporting the shutters from shop to house, or house to garage, is very simple. You can lift five or ten at one time.

## GREAT APPEAL OF A THERMAX PLATE

Thermax, described in detail in Chapter 7, is near-ideal as a material for an indoor shutter. Produced by Celotex Corp., a subsidiary of Jim Walter Corp., it consists mainly of a slab of isocyanurate foam that contains a small amount of fiberglass which adds to the strength and stiffness. Bonded to each face is a sheet of 0.002-in. shiny aluminum foil. The R-value of 1-in. Thermax is about 8. Standard thicknesses (in inches) are: 1/2, 5/8, 3/4, 7/8, 1, 1¼, 1½, 1¾, 2. Standard plate size is 8 ft × 4 ft. Density: about 2½ lb/ft$^3$. Retail cost of 1/2-in. Thermax: 26¢/ft$^2$.

The aluminum foil is helpful in many ways:

It helps seal in the high-molecular-weight foamant gas, which has especially low thermal conductivity.

It forms a smooth, attractive surface to which little dust clings; the surface is easily wiped clean.

It helps protect the foam from nearby flames. But the protection is minor, and may be extremely brief.

It excludes moisture, small bugs, etc.

It may be painted with almost any kind of paint. Decorations are easily applied.

Sticky tape adheres to it well, and the aluminum foil tends to distribute forces over a large area of the foam. (The foam itself is easily punctured or torn by highly localized forces.)

It reflects visual range, near-IR, and far-IR radiation; thus it can be used as a solar-radiation reflector. Its role in reducing heat-loss by reflecting far-IR is minor inasmuch as the isocyanurate foam itself is highly effective with respect to conductive flow and radiation.

Thermax plates that I have used as shutters showed, after 2 months' use, no sign of warping.

The material can be cut in a few seconds with a carpenter's saw or a sharp knife.

Exposed edges should be covered with tape. Otherwise the edges are abrasive to the skin and they may suffer mechanical damage. Nearly any kind of sticky tape can be used.

Warning: The material is flammable and the products of combustion are poisonous. It should not be used where there is appreciable danger of fire or where prohibited by fire regulations.

In making and using not-permanently-attached shutters of 1/2-in. Thermax, I have found that:

No edging (other than tape) is needed. No wooden or plastic strips are needed. I used no frame at all—just the sheet of Thermax.

No sheathing is needed: no plywood or Masonite or the like.

No stiffeners or diagonal braces are needed.

Because no edging is used, and because the material is so easily cut, securing close tolerance on fit is easy. One can start with a plate that is generously large, try it out, and then slice off as much material as is necessary. No careful measuring is needed.

Attaching small fittings is easy. Hinges and knobs can be attached in seconds, as explained in a later section.

If a plate threatens to interfere with (strike) a window latch, for example, a suitable notch can be cut in the plate with a knife so that no interference can occur.

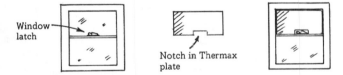

Window latch

Notch in Thermax plate

### APPEAL OF A STYROFOAM PLATE

The properties of Dow-Corning Styrofoam are listed in Chapter 7.

In many ways the material is like Thermax: high R-value (not quite so high), easily cut with a saw or sharp knife, flammable. Thicknesses from 1 in. to several inches are available. Typical sheet width: 24 in. One of the commonest kinds of Styrofoam is Styrofoam SM, which is light blue in color.

The material is not as strong or stiff as Thermax. Therefore shutter builders will probably use, ordinarily, sheets 1-in. or 1½-in. thick. I have made many shutters of 1½-in. material, and they seemed strong and remained flat.

Using thick sheets (1½ in. or more), one can make butt joints merely by applying sticky tape to adjoining faces. Thus a shutter builder can make big plates out of little ones, as well as vice versa.

When the material is cut with a saw in a low-humidity environment, electrostatic effects are large and the resulting "sawdust" tends to adhere to almost everything: saw, hands, clothing, workbench, broom, etc. Doing the sawing outdoors on a lawn may eliminate the need for a difficult sweep up.

The edges of a Styrofoam SM sheet are not abrasive. Thus there is no need to apply tape to them.

The faces are not covered with aluminum foil. Therefore they are not highly reflective. Also, the material is damaged by various common kinds of paints. Latex paints, however, do no damage.

If the Styrofoam plate used is thick (1½ in. or more), it may be grooved to "accept" window muntins. If the muntins project 3/4 in. from the window glass, and if a 1-in. groove is made in the Styrofoam SM (with a sharp knife), the plate can be pressed firmly against the glass, bridging the muntins. This procedure permits a face-seal despite the muntins.

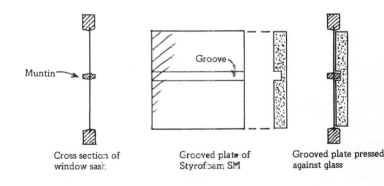

Cross section of window sash.

Grooved plate of Styrofoam SM

Grooved plate pressed against glass

## PLATES OF OTHER MATERIALS

Other materials that may be used include:

Urethane foam
Beadboard
Thermoply
Plywood
Regular wood
Celotex
Corrugated cardboard.

Although lacking the appeal of Thermax, these materials may sometimes be chosen because of other special properties. Beadboard is extremely cheap. Wood is very strong.

Corrugated cardboard can be obtained free— at the back doors of stores and supermarkets. If three sheets of corrugated cardboard are pasted together and a wrapping of aluminum foil is applied, one has a high-performance shutter at trivial cost. In crude tests, I found that such a shutter, applied to a double-glazed window, reduced heat-loss by 50%.

## PARKING PROBLEM

Most of the not-permanently-attached shutters in a house are removed by the residents each morning to admit solar radiation and provide illumination and view. In little-used rooms, some or all of the shutters may be left on the windows all day, especially if the shutters have small central cut-out areas (peep holes) such as are discussed on page 62.

Often, a shutter that is removed may simply be set on the floor, leaning against the wall.

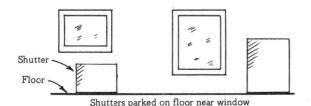

Shutters parked on floor near window

Another strategy, useful if there are many windows fairly close together, is to park the shut-

ters in a bunch, preferably in a not-very-conspicuous place. Any of these procedures may be used:

● In each room, keep one window (preferably *not* a south window) permanently shuttered. Install all the other shutters on this window; that is, make one big stack of shutters here. Make the window sill extra wide and provide a raised rim that will prevent the shutters from sliding off. A restraining clip at the top may or may not be needed.

● Stack the shutters on a special shelf (with a rim) on an unused area of wall.

● Make a 2-inch-diameter hole near the top of each shutter and hang all the shutters from one horizontal, 1-inch-diameter, 1-foot-long rod projecting out from a wall. If the rod is at least 7 ft above floor level no one will ever strike his head on it. Note:  A small hole near the top of a shutter does not significantly impair its ability to reduce heat-loss.

● Rest the shutters on the floor or lean them against a wall. Choose a site behind a big chair or bookcase that is fairly close to the wall.

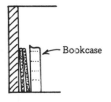

Bookcase

● Slip the shutters behind a door which, when open, is parallel to a wall and about a foot from it.

Door

Plan

● Provide a section of false wall about a foot from the main wall. Store the shutters in the intervening space.

False wall

Plan

● Store the shutters horizontally just below the horizontal top of a large table or counter. Provide special support brackets.

Table

Bracket

● Store the shutters in a closet or a hallway or under a cellar staircase.

● If the ceiling is high, store the shutters on a large platform attached to wall or ceiling. Pushing

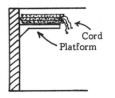

Cord
Platform

the shutters into place there is easy, even for a fairly short person. But a short person may have difficulty retrieving the shutters unless a short length of cord is attached to the end of each shutter and hangs down.

● Store the shutters on a small, lightweight cart, which is moved, during the day, to an out-of-the-way location. Such a scheme has been proposed by D. A. Block and L. Hodges (see Ref. I-402f, p. 771).

Complications arise if, to suit windows of different sizes and shapes, the shutters are of slightly different sizes and shapes. When the residents are preparing to reinstall the shutters, how will they know which belongs where? The shutters should be clearly labeled, or color coded.

In summer the shutters may be stored in a basement, garage, or attic.

## KNOBS AND HANDLES

Here are some ways of applying knobs and handles to a shutter that consists of a simple removable plate:

Make a knob by bending up a 1-inch-wide strip of sheet metal (aluminum, copper, or other metal) that is about 0.025 in. thick. Provide large terminal flanges. Tape these to the plate, using

1"

Knob (bent strip of metal)

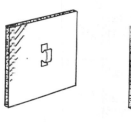

One large central knob

Two small knobs

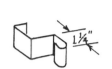

Clip-type knob

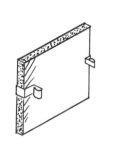

Two clip-type knobs,
one at each edge

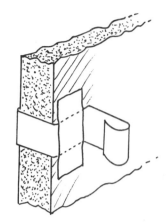

Clip-type knob attached
by tape

almost any kind of sticky tape, e.g., masking tape, duct tape. Install one large knob at the center of the insulating plate or install two small ones, one at each side. The less the knob protrudes, the more compactly a set of shutters can be stored. A projection of about 1 in. may be ideal.

Make a clip-type knob that is designed to embrace the edge of the insulating plate. Apply tape at the front and back of the plate.

Employ a 1-inch-wide strip of heavy cloth tape. Allow a 5-in. length to hang down on the room-side of the shutter, for use as a handle. Pass the other end of the tape through a slot-like hole in the shutter and attach it at the back with tape. Use two such handles, one at each side. When several shutters are stacked, these handles take up practically no room.

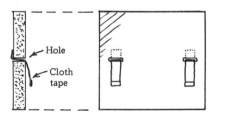

If the plate is not tough and there is danger that the tape will tear its way through it, employ a "bed-strip" of sheet metal on the back face, as indicated in the sketch.

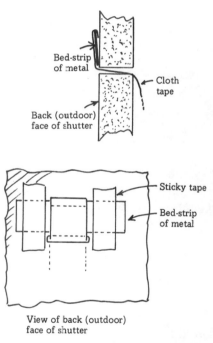

View of back (outdoor)
face of shutter

If the shutter includes wooden framing strips, almost any kind of handle or knob can be attached to the strips with screws. However, applying screws directly to Thermax, Styrofoam, or the like is usually not effective: the screws get very little grip and pull out easily.

## SEALS

See Chapters 4 and 21 and Appendix 2.

## INDOOR SIMPLE PLATE THAT IS PRESSED AGAINST WINDOW AND RESTS ON SILL (Scheme 13.1)

Here I describe many ways of mounting a single rigid plate of Thermax, Styrofoam SM, or the like on the simplest window. The plate is pressed against the window (against the sash or against the glass itself) and rests on the sill. Because the plate is lightweight, little force is needed to hold it in place. And because it presses against the glass or sash, providing a face seal, tight fit is not needed, as explained in Appendix 2. Another helpful fact is that a small gap at the top, e.g., a long horizontal gap 1/2-in. or 1-in. high, does not significantly impair the performance.

There are so many ways of securing such a plate to a simple window. One can use friction, buttons, retaining strips, ears, clips, prongs, magnets—and even potted plants!

### Scheme 13.1a

Use friction in the simplest way. Merely cut the plate to just the right size and press it into the aperture of the fixed frame of the window. Press it inward (south) until it bears against the sash, or, if there is no sash, presses against the glass itself. Provide two small handles such as have been described on a previous page; they are needed for removing the plate in the morning.

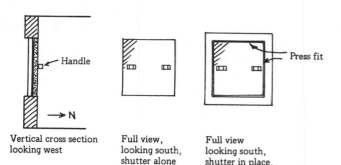

Vertical cross section looking west

Full view, looking south, shutter alone

Full view looking south, shutter in place, in use

### Comment

In removing the plate, the resident cannot go wrong. It can be removed all at once, or can be pulled away bottom first and then the top, or vice versa. No matter how one goes about removing it, no harm results.

Because a small gap at the top does no significant harm, it probably does not pay to devote effort to achieving tight fit there.

### Limitations

If the plate shrinks, it will no longer be held by friction and may fall off.

If it expands, it will not fit into the space in question.

If different windows have slightly different widths, the plates too must have different widths. They will not be interchangeable.

### Scheme 13.1b

Make the plates slightly undersize and build up one vertical edge with tape (i.e., increase the width by tiny increments) until a good press fit is achieved. If the plate shrinks, add more tape. (Alternatively, the tape could be added to the top of the plate.)

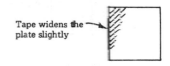

Tape widens the plate slightly

### Scheme 13.1c

As above, except apply the tape to the window jambs. Adjust each window aperture this way so that all will have the identical inside width and thus any plate may be installed on any window.

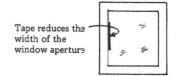

Tape reduces the width of the window aperture

### Scheme 13.1d

Add a compressible strip to one side of the plate. If the strip can be compressed 1/8 in., a given plate can accommodate as much as 1/8 in. variation in inside width of window frame.

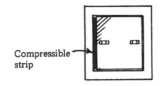

Compressible strip

The compressible strip (tacked, cemented, or taped to edge of the plate) may have a rectangular cross section, or a cross section that is circular, U-shaped, or L-shaped.

Vertical cross section of plate and compressible strip

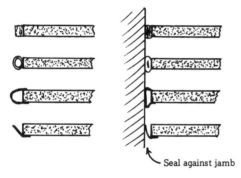

Seal against jamb

### Scheme 13.1e

Use buttons that turn. Use two buttons if the plate is stiff and flat, and use four if it is flexible and may warp. The buttons may be at the sides and/or top or bottom.

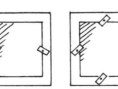

Simple button     Beveled button

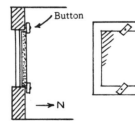

Button

N

Various arrangements of buttons

Note: The buttons should be short: they should overlap the plate only about 1/3 in. If the overlap is much greater, damage may be done to them or to the plate if the plate is pulled away from the window while one button is still engaging it.

If the window is so deeply recessed or the plate is so thin that the north face of the plate is not flush with the face of the fixed frame of the window, one can "fatten" the plate with battens. The battens can be taped or cemented to the plate.

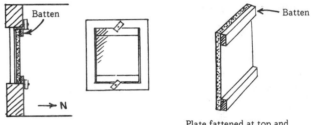

Plate fattened at top and bottom with battens

### Scheme 13.1f

Use a restraining (capturing) strip at the top of the fixed frame of the window. To install the plate, slip the upper edge under the restraining strip, press the lower edge against the window, and allow the plate to rest on the sill. Friction keeps the lower edge in place on the sill.

The height of the plate is slightly less than that of the window aperture. Thus when the plate is in place, there is a small gap at the top—which does no significant harm, as explained in Appendix 2. The restraining strip should overlap the plate by only about 1/3 in. to insure that no severe stress is imposed on the strip or the plate when the plate is being removed.

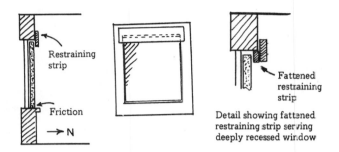

Detail showing fattened
restraining strip serving
deeply recessed window

If there is danger that the lower edge of the plate will move away from the window, tiny ridges or bumps can be provided on the sill to discourage such motion. The ridges may be only 1/8-in. high and may consist of projecting heads of round-head screws, tacks, or double-pointed tacks. Inverted U-shaped pieces of heavy wire may also be used. See accompanying sketches.

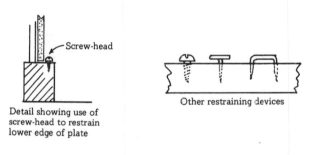

Detail showing use of
screw-head to restrain
lower edge of plate

Other restraining devices

### Scheme 13.1g

Instead of a restraining strip, use ears, tabs, or clips. A restraining strip is simple to make and install and has a simple, clean appearance. But functionally it constitutes overkill: it has vastly greater area and greater strength than is needed merely to hold the upper edge of the plate in place. Various schemes using ears, tabs, and clips are depicted in the accompanying sketches.

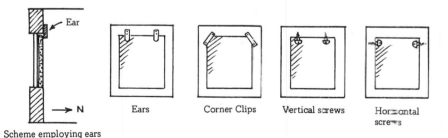

Scheme employing ears

Ears

Corner Clips

Vertical screws

Horizontal screws

### Scheme 13.1h

Use projecting prongs at the top or bottom of the plate. They fit into small, 1/4-inch-deep holes in the fixed frame of the window. If the plate itself is not strong enough to hold the prongs, one may use a special kind of taped-on clip of sheet metal that includes a prong.

I have little enthusiasm for prongs because (1) they are hard to apply to foam-type plates; (2) using them entails drilling holes in the fixed frame of the window, and (3) most of the schemes discussed immediately above seem superior.

One can devise retractable prongs. One can even devise an opposing pair of prongs (say, one at left and one at right) that can be retracted merely by turning one central control knob.

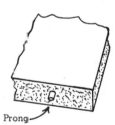

Detail of prongs at bottom
and compressible strip at top

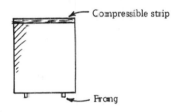

Detail showing prong on bottom
edge of plate

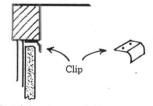

Detail showing use of thin
metallic clip attached to
upper inner part of fixed
frame of window

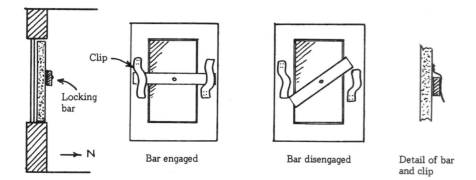

Bar engaged          Bar disengaged          Detail of bar
and clip

### Scheme 13.1i

Use a central transverse locking bar. The plate is pushed into place against the window and the bar is then turned so that each end engages a fixed clip. Note: To attach the bar to a relatively soft foam-type plate may be difficult.

### Scheme 13.1j

Hold the plate in place merely by means of a suspended weight that leans against the plate. For example, obtain a 10-lb potted plant 8 in. in diameter, and suspend it so that its edge tends to press lightly against the window. To install the insulating plate, pull the potted plant a few inches away from the window with your left hand and insert the plate with your right hand. Then release the potted plant, which will swing against the plate and maintain a steady force of about 1 lb on it.

Instead of a potted plant, one may use almost any other decorative item that has comparable weight and diameter, such as a one-gallon plastic bottle or jug filled with water (or colored water), a lobsterman's buoy, a heavy birdcage, a pumpkin, a

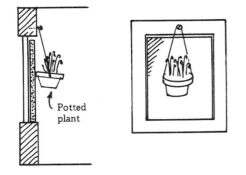

heavy lamp, a water-filled copper tea kettle, a pair of ski boots.

## INDOOR SIMPLE PLATE SUPPORTED BY THE EDGES OF THE SASH (Scheme 13.2)

Here the plate does not rest on the sill. It is suspended from fixed prongs integral with the inner edges of the sash. Furthermore, it is cut to fit within the sash frame; that is, it is pressed directly against the glass.

Assuming the plate to consist of 1-in. Thermax, one prepares two special clips, each with a sloping notch and an integral handle as indicated in the sketches. Each clip is made of sheet aluminum (or steel, copper, or brass) that is just thin enough (about 0.025 in.) to be fairly easily bent into shape with the aid of a steel vice and a hammer. The metal can be cut with shears, and edges and corners are rounded with a file. Each clip has a span of 1 in., and one end is bent around to form an integral handle. The most important part of the clip is a sloping notch. The clips are attached to the upper portions of the left and right edges of the plate by means of tape.

A 1-in. round-head screw is installed in the inner portion of the left member of the sash frame. The screw is horizontal and is 3/8 in. from the glass. It is about 3 in. below the upper edge of the sash. About 1/3 in. of the screw projects, i.e., is exposed. A similar screw is installed in the right member of the sash frame at the same height.

To install the plate, one grasps it by the handles and presses it into place in such a way that the sloping notches engage the horizontal screws.

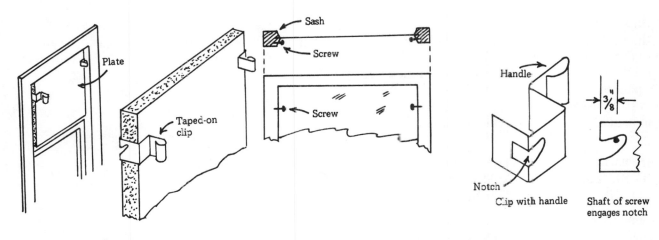

Clip with handle    Shaft of screw engages notch

Because the slots are sloping, gravity urges the plate toward the glass, and since the slots are nearer the south face of the plate than the north face, pendulum effect urges the lower portion of the plate toward the glass. In summary, all portions of the plate are urged toward the glass and accordingly an excellent face seal is achieved automatically. This is true despite the fact that no precision is required in making or installing the plate. Gravity does the trick, making the plate slide southward (as permitted by the notches) until a good seal is achieved. (In fact, the seal is needlessly good; even if there were a 1/4-in. airspace between glass and plate the seal would still be highly effective.)

Note that the screws are, in a sense, recessed. Thus they do not interfere with opening or closing either sash.

This scheme was invented and tried out by me late in 1978. Performance was excellent even though construction was slipshod. I found that the worst mistake I could make was mounting one screw higher than the other; but this could be compensated for immediately by relocating one of the clips slightly higher or lower on the plate. I called this scheme S-3/27/79, this being the date of a formal report on the subject.

I found that, instead of using the two screws mentioned, I could use two taped-on strips of sheet metal bent and shaped to engage the notches. However, screws seemed cleaner and more durable.

### Scheme 13.2a

Relocate the horizontal supporting screws so that they extend northward (or slightly upward and northward) from the side members of the sash.

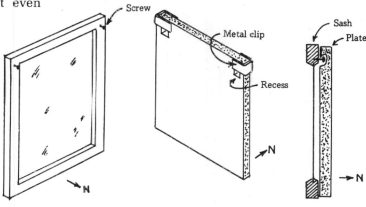

Sash, perspective view    Plate, perspective view    Plate mounted on sash, vertical cross section

Provide metal-edged recesses on the south face of the plate. (Installing the screws on the north faces of the sash members is easier than installing them on the inner edges.) If the screws are situated high up on the upper sash of a double-hung window, it is still possible to raise the lower sash a considerable distance before interference occurs.

### INDOOR SIMPLE PLATE SUPPORTED BY BEVELED PLASTIC STRIPS CEMENTED TO THE GLASS ITSELF (Scheme 13.3)

Cement two beveled strips of thick transparent plastic (e.g., 1/4-in. Plexiglas or Lucite) to the window, and provide metal-edged recesses in the insulating plate. To install the plate, press it against the glass sheet and lower it (the plate) a fraction of an inch so that the metal pieces will engage the beveled plastic strips and produce a good face seal.

When the plate is not present, the plastic strips remain but are almost invisible.

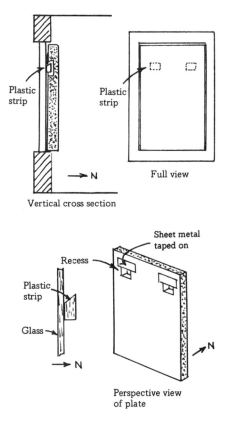

Plastic strip

Plastic strip

Full view

Vertical cross section

Sheet metal taped on

Recess

Plastic strip

Glass

N

N

Perspective view of plate

If there is a chance that the taped-on pieces of sheet metal will come loose, they can be made extra long and can be wrapped around the edges of the plate.

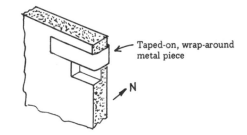

Taped-on, wrap-around metal piece

N

### INDOOR SIMPLE PLATE AFFIXED TO THE GLASS BY MAGNETS (Scheme 13.4)

This scheme, developed several years ago by S. C. Baer of Zomeworks Corp., makes much use of pairs of magnetically attracting strips which have the trademarked name *Nightwall Clips*. A typical pair consists of (1) a permanent magnet 3 in. × 1/2 in. × 1/16 in. and (2) a piece of ordinary galvanized steel 0.025-in. thick (24 gauge). Each has a pressure-sensitive adhesive backing that is virtually unaffected by prolonged exposure to solar radiation.

Such magnetic pairs are used to secure the insulating plate to the glass sheet of the window. The pairs are situated along the edges of the plate and edges of the glass sheet and are about 18 to 24 in. apart on centers. The galvanized pieces are attached to the glass and the magnets are attached to the plate. The adhesive backings integral with the pieces provide a strong bond if the pieces are pressed home strongly.

To insure that the locations of the steel pieces on the glass exactly match the locations of the

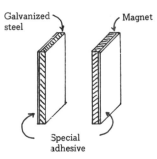

Galvanized steel

Magnet

Special adhesive

magnets on the plate, install the strips as pairs on the plate, then remove the inert backings from the exposed surfaces of the pairs and press the entire plate plus array of pairs against the glass, causing all of the galvanized pieces to adhere permanently to the glass in exactly the correct positions.

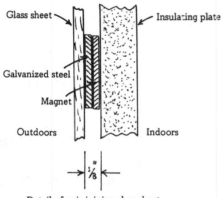

Detail of pair joining glass sheet
and insulating plate (not to scale)

The 1/8-in. airspace between the glass and the plate does not appreciably affect the thermal performance of the insulating plate. (One could, of course, recess the pairs into the plate so that the airspace in question would be much less than 1/8 in.)

To open shutter: grasp plate and pull.

Cost of pair: About 35¢, from Zomeworks Corp.

## Comment

This scheme is simple, very inexpensive, and performs well. Installing the clips takes only a few minutes. No tools are needed.

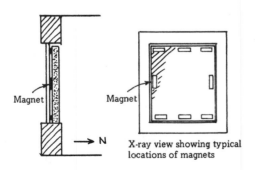

X-ray view showing typical
locations of magnets

## INDOOR FOLDING SHUTTER THAT IS INSTALLED SYMMETRICALLY (Scheme 13.5)

This shutter, which can be folded and then parked in a small space, consists of two rigid, lightweight plates joined by a hinge: a flexible tape extending the full width of the shutter.

Two fixed ears are provided at the bottom of the fixed frame of the window, to hold the bottom of the shutter, and two long, springy tabs are provided at the top to urge the upper plate (and indirectly, via the hinge, the lower plate) toward the window.

To open the shutter: Grasp the handles (not shown) situated near the middle of the shutter and pull. Remove, fold, and store the shutter.

To close the shutter: Pick up the shutter, open it up to a 90-degree angle, and press it against the window in such manner that the upper part slides upward and the lower part slides downward.

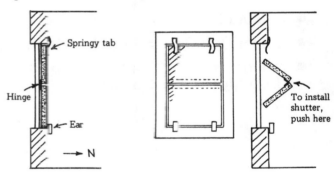

Scheme 13.5a

Dispense with the springy tabs at the top—because they are expensive and visually obtrusive. Use fixed ears at the top and bottom and buttons at the sides.

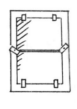

Scheme 13.5b

Use fixed ears at the top and bottom and employ a hanging potted plant to press the center of the shutter toward the glass

# CHAPTER 14

# INDOOR OPAQUE SHUTTERS THAT SLIDE

- Indoor Plate That Slides (Without the Aid of Rollers) in Horizontal Channels (Scheme 14.1)
- Indoor Plate That Slides (With the Aid of Rollers) in Horizontal Channels (Scheme 14.2)
- Indoor Plate That Slides in Vertical Channels (Scheme 14.3)
- Indoor Quilt That Slides Over the Top of a Trombe Wall (Scheme 14.4)
- Indoor Train of Insulating Boards That Slides Over the Top of a Trombe Wall (Insulider Shutter) (Scheme 14.5)

Sliding shutters can slide horizontally (right and left) or vertically (up and down). The former, being more common, are discussed first.

## INDOOR PLATE THAT SLIDES (WITHOUT THE AID OF ROLLERS) IN HORIZONTAL CHANNELS (Scheme 14.1)

Such a shutter was installed about 200 years ago in the Sarah Orne Jewett house in South Berwick, Maine, and is still in use. The shutter is of wood 1-in. thick, and rests in a 1/2-in. channel in the wooden sill. To open the shutter, one slides it laterally into a slot (about 1.1-in. thick) in the adjacent wall. There are no rollers.

A few years ago M. Jantzen of Carlyle, Ill., designed and built a sliding shutter consisting of a plate of Technifoam edged with wooden strips 2 in. × 2 in. in cross section. The horizontal channel was made of aluminum. Seals along the edges were provided by rubber strips.

A somewhat similar shutter was developed a few years ago by Aardvark and Sons of West Yarmouth, Mass. A patent application for their specific design was filed about 1978. Their device employs no channel. It rests directly on the flat sill and is confined by an upward-projecting guide strip attached to the north edge of the sill. The accompanying sketches show the design.

### Limitations

The shutter must be planned before the wall is built.

It requires special construction and a very thick wall.

The seal may or may not be sufficiently tight.

Question: Could one get away with a shutter consisting just of a thick Thermax plate, the edges of which have been covered with a tough pressure-sensitive tape that has a low coefficient of friction? Could one skip the wooden frame? Is the Thermax plate light enough so that it can slide hundreds of times on a tape edging? Would an aluminum tape meet the requirements?

## INDOOR PLATE THAT SLIDES (WITH THE AID OF ROLLERS) IN HORIZONTAL CHANNELS (Scheme 14.2)

R. DeWitt of Isothermics Co. has designed such a shutter, the main component of which is a sheet of 1-in. Thermax sheathed with 1/8-in. plywood and edged with strips of 1-inch-thick pine wood. At the top of the sheet are two pairs of rollers (Lawrence Brothers, Inc., Hanger #582) which run in a horizontal track (Lawrence #580). The bottom hangs free. There is no channel or other guide here.

When the shutter is not needed, it is slid into a slot defined by a sheet of 1/2-in. plywood.

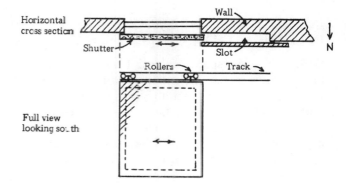

Horizontal cross section
Wall
Shutter
Slot
N
Rollers
Track
Full view looking south

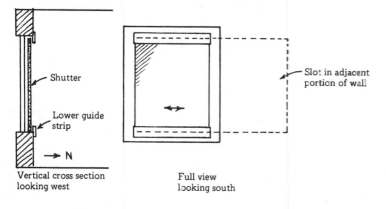

Shutter
Lower guide strip
N
Slot in adjacent portion of wall
Vertical cross section looking west
Full view looking south

*Scheme 14.2a*

A somewhat similar design was developed by John M. Newman of St. Louis, Mo. The plate consists of a hollow-core door. The lower edge of the plate is confined by a horizontal strip of wood. When the plate is slid to open position, it remains in full view.

*Scheme 14.2b*

A shutter developed by Kingston and Sons of Stockbridge, Mass., is somewhat similar to the above-described shutter. The Kingston shutter is faced with Homosote or Sheetrock, has a wooden frame, and is filled with urea formaldehyde foam. It is guided at top and bottom by ball bearing rollers that run in tracks of metal or wood. Edge seals are provided by strips of Velveteen or Velcro.

When such a shutter is installed in a new house, a slot is provided to receive and conceal the shutter when it is not in use. In retrofit installations no slot is provided.

## INDOOR PLATE THAT SLIDES IN VERTICAL CHANNELS (Scheme 14.3)

D. S. Kelbaugh uses a shutter in his Princeton, N.J., house that consists of a 1-in. plate of Styrofoam wrapped in cloth. Kelbaugh reported: "It slides up and down in aluminum channels and, when open, is held up by friction alone—very easy to operate and pretty air-tight." Cloth-tape handles are attached to the upper edge. A fixed wooden strip attached to the sill serves as a seat for the closed plate.

A somewhat similar scheme has been developed by R. S. Levine of the University of Kentucky.

His window is about 20 in. × 20 in. and is mounted between 6-in. studs. The insulating plate is of 1½-in. Thermax. When it is slid to "up" position, it is held there by two pegs which are attached to the lower edge of the plate by strings and serve also as down-pulls.

### Limitation

The scheme is applicable only if the window is small and if there is much free space above it.

*Scheme 14.3a*

As above, except extend the channels downward, rather than upward, and lower the plate to open it.

## INDOOR QUILT THAT SLIDES OVER THE TOP OF A TROMBE WALL (Scheme 14.4)

At night the south face of a Trombe wall is insulated by a thick flexible quilt. At the start of the day the quilt is slid over the top of the wall and lies close to the north side. This scheme has been described in detail in my *Inventions* book (S-235cc).

### Comment

Although extremely simple, the system is effective and versatile. It greatly cuts nighttime heat-loss from the Trombe wall to the outdoors, and during the day it cuts heat-flow from wall to room—helpful inasmuch as the room is likely to be warm enough throughout the day, especially if there is some direct passive solar heating also. If the resident wishes to have both faces of the wall uncovered, he may remove the quilt entirely or may "heap it up compactly" on the top of the wall.

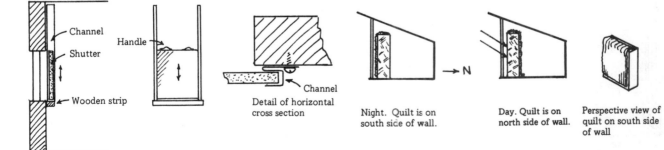

Channel
Handle
Shutter
Wooden strip
Channel
Detail of horizontal cross section

Night. Quilt is on south side of wall.    Day. Quilt is on north side of wall.    Perspective view of quilt on south side of wall

## Limitations

The quilt does not make firm contact with the face of the wall.

Sliding the quilt over the top of the wall may be difficult if the wall has sharp edges.

If the wall is very tall, the resident may sometimes find the quilt edge too high to reach.

The quilt may occasionally fall entirely off the wall.

*Scheme 14.4a*

As above, except:

Make the wall tapered: thin at the top, thick at the bottom. Then gravity will ensure that the quilt lies very close to the face of the wall.

Mount slender rollers along the upper edges of the wall.

Attach cords to the four corners of the quilt and fasten the lower ends of the cords to the base of the wall. Then the corners of the quilt are effectively always within reach and the quilt can never slide entirely off the wall.

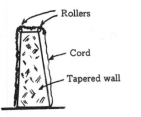

## INDOOR TRAIN OF INSULATING BOARDS THAT SLIDES OVER THE TOP OF A TROMBE WALL (INSULIDER SHUTTER) (Scheme 14.5)

This articulated shutter, developed by Jaksha Solar Systems and called *Insulider*, is much like the one described on p. 99 and used in between-glazings location. Here we deal with a different location (indoor location) and a special application: to the space between the glazing and the Trombe wall.

The heart of the device is a set, or train, of contiguous insulating boards—for example, a set of Styrofoam boards 1½-in. thick and 3-in. wide—

permanently attached to a tough, flexible, slippery backing sheet, e.g., a fine-weave, rip-stop nylon cloth. If the Trombe wall is 8-ft long and 6-ft high, the above-described system (shutter) is 8-ft wide and 6-ft high. It can be slid upward to expose the Trombe wall to solar radiation, or can be slid downward, between the wall and the window, to prevent flow of heat from the wall to the window. When sliding along, the assembly turns a corner, i.e., bends. At the lower edge of the shutter there is a heavy horizontal bar which helps the shutter descend.

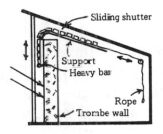

When the shutter is slid upward, with the aid of ropes running over pulleys, it turns and slides parallel to the ceiling and is guided and supported by a curved platform of sheet aluminum or other slippery sheet. The ropes are attached to the nylon-cloth backing sheet via a batten which distributes the force so that the risk of tearing the nylon cloth is avoided.

Such a shutter, besides being applicable to a Trombe wall, may serve a drumwall or other heat-storing wall. Also it can be applied to greenhouse windows or large windows of dwellings.

### Comment

The device cuts heat-loss greatly, is easy to operate, and may be nearly foolproof.

# CHAPTER 15

# INDOOR OPAQUE SHUTTERS HINGED AT SIDE

- How to Attach Hinges to a Plate of Rigid Foam
- Indoor Two-Plate Shutter for Deeply Recessed Window: The Two Plates are Held at the Sides by Hinges That Are Close to the Glazing (Scheme 15.1)
- Indoor Automatic Two-Plate Shutter for Deeply Recessed Window: The Two Plates Are Attached by Vertical Pivots One Foot North of the Glazing (Levine Shutter) (Scheme 15.2)
- Indoor Two-Plate Shutter Employing Thermoply (SUN SAVER Shutter) (Scheme 15.3)
- Indoor Two-Plate Shutter Employing Rigid Foam Core (Scheme 15.4)
- Indoor Shutter Employing Two Pairs of Hinged Folding Plates for Deeply Recessed Window (Scheme 15.5)
- Indoor Shutter Employing Two Pairs of Hinged Folding Plates for Window That Is Only Slightly Recessed (Scheme 15.6)

Here I describe schemes that are somewhat more complicated conceptually but which have certain advantages, such as simplifying the problem of parking the shutter when it is not in use. In each case the plates are attached by hinges at the sides, and the plates can be swung to right or left.

Indoor shutters that are hinged at the top are discussed in the following chapter.

Indoor shutters that are hinged at the bottom have a great disadvantage: if opened wide, they project far out into the room and may be struck by anyone walking past. (However, see Chapter 19 for descriptions of some very special, partially opening shutters that are hinged at the bottom.)

Most shutters that are attached by hinges have strong wooden frames and many have faces of plywood or the equivalent. The hinges are attached to the pertinent frame member.

Where a large-area shutter is required, hollow-core wooden doors can sometimes be used successfully. (I am indebted to Prof. Fuller Moore for pointing this out.)

Could one make a shutter out of rigid foam (Thermax, or Styrofoam, e.g.) as is—and attach hinges to it? This subject is discussed in the next section.

## HOW TO ATTACH HINGES TO A PLATE OF RIGID FOAM

Because screws and nails get little or no grip on rigid foam plates of Thermax, Styrofoam, or the like, attaching hinges is a problem.

Some possible solutions are sketched below.

● Take any ordinary hinge and glue a 5 in. × 2 in. piece of 0.025-in. steel (or copper or aluminum) to the half of the hinge that is to swing. Attach this piece of steel to the rigid foam plate by means of pressure-sensitive tape that covers all of the exposed edges of the piece of steel and also wraps around the adjacent edges of the foam plate.

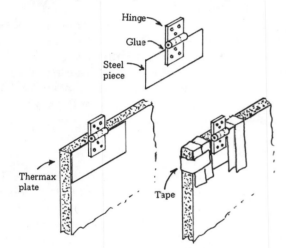

● As above, except make a U-shaped piece of steel, glue one face of it to the movable half of the hinge, insert one edge of the foam plate in the "U," and secure it with tape applied along the edges of the "U."

● Alternatively, one may employ a pivot. This engages a tab that is integral with a U-shaped clip. (See also p. 69.)

## INDOOR TWO-PLATE SHUTTER FOR DEEPLY RECESSED WINDOW: THE TWO PLATES ARE HELD AT THE SIDES BY HINGES THAT ARE CLOSE TO THE GLAZING (Scheme 15.1)

This scheme, applicable only if the window is narrow and deeply recessed, was suggested to me on 8/3/77 by C. La Porta.

The two rigid insulating plates are supported by hinges that are mounted close to the glazing. When opened (swung through 90 degrees), the plates lie close to the jambs.

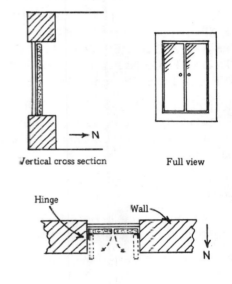

Vertical cross section          Full view

Horizontal cross section

### Limitations

When closed, the plates may not be sealed well.

If the window is wide, the two plates also must be wide and, when open, will project far into the room.

### Scheme 15.1a

As above, but employ buttons or small magnets to hold the plates tightly closed.

### Scheme 15.1b

Here I assume that the window jambs, or reveals, flare steeply outward toward the room. The open plates, likewise, may flare outward, which greatly reduces their tendency to project far out into the room.

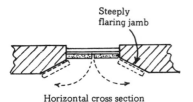

Horizontal cross section

## INDOOR AUTOMATIC TWO-PLATE SHUTTER FOR DEEPLY RECESSED WINDOW: THE TWO PLATES ARE ATTACHED BY VERTICAL PIVOTS ONE FOOT NORTH OF THE GLAZING (LEVINE SHUTTER) (Scheme 15.2)

This device, invented in about 1975 by R. S. Levine of the University of Kentucky and applicable to narrow, tall, deeply recessed windows, has these unusual features:

It is operated automatically by a solar-powered opener-closer (actuator).

In one special mode of operation it serves as a main part of an air-type collector.

On hot summer days it may serve as part of a solar-powered chimney for use in venting hot air and bringing in new air.

The shutter includes two 1¼-inch-thick Thermax plates. If the window is about 2-ft wide and is recessed about 1 ft, each plate is about 1-ft wide and is supported by pivots at one edge which is about 1 ft from the window. When the plates are closed they lie in a plane about 1 ft from the window and, together with the jambs, form a box, or conduit, about 2 ft × 1 ft in horizontal cross section. Seals are provided by compressible strips. The south faces of the plates are painted black.

Each plate is opened or closed by means of a slender belt that runs on two pulleys, the axles of which are vertical. The larger of the pulleys (master pulley) can be rotated about 1/5 turn by a horizontal rod that forms a main part of a solar-powered (thermally powered) opener-closer, an actuator that costs about $20 and is made by Dalen Products, Inc. Nearly always the rod is either fully extended (plates closed) or fully retracted (plates open). The main pulley and the actuator serve both plates. One of the belts has a full twist and the other has no twist; accordingly when one panel turns clockwise the other turns counterclockwise. The entire control system (actuator, pulleys, belts) is situated near the bottom of the window.

### Operation

When the actuator is warmed (e.g., by solar radiation) and becomes hotter than about 70°F, wax in the actuator melts and increases in volume, extending the actuator rod and turning the pulley about 45 degrees. This causes the belts to turn the small pulleys—and shutters—90 degrees, to open position. When the sun sets and the actuator cools, the operation is reversed.

Overriding manual controls are provided also.

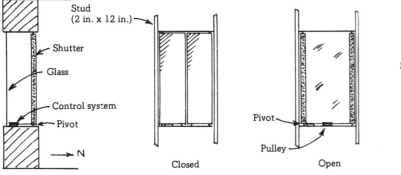

Full view looking south

Horizontal cross section
shutter closed

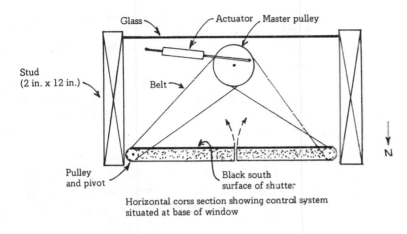

Horizontal corss section showing control system situated at base of window

## Additional uses

Such window-and-shutter systems (called by Levine "Sundows") can be put to important additional uses, as he has demonstrated in his Raven Run house in Lexington, Kentucky. In this house, which has about 100 windows, there are several sets of systems (such as have been described above) one above another, in vertical series, with clear air passages from one to the next. These series can be used as follows:

Sunny day in winter: Close the shutters, so that much solar energy is collected in the duct-like space between glass and shutter. Arrange for the hot air emerging from the top of the series to flow to a north room or to a storage system. Allow cool air near the floor to enter the bottom of the series. To prevent reverse flow at night, dampers are provided. Some windows may be used in this manner to provide heat to be stored and used later, while others may be used in the usual way to provide direct passive gain.

Hot sunny day in summer: Proceed as above (use a connected vertical series of windows and shutters) except vent the hot air (air issuing from the top of the series) to the outdoors. This helps keep the house cool.

Notice the important, money-saving consequences of using studs that are 12 in. wide and 24 in. apart. The shutters can be installed between the studs, and the pivots can be mounted so far (about 1 ft) from the glass that, when the shutters are closed, they define a duct of generous cross section. That is, the glass, studs, and shutter plates form a complete, nearly-air-tight duct—ready for use as a major part of an air-type collection system in winter and part of a passive ventilation system in summer. In a sense the duct is free; the components mentioned do double duty.

### Limitations

The 12-inch-thick wall takes up much space.

The 12-inch-wide studs are expensive.

Because the studs are so wide and are only 2 ft apart on centers, (a) view off to the sides is limited and (b) direct sunlight can proceed directly into the room only from about 8:00 or 9:00 a.m. until about 3:00 or 4:00 p.m.

When the shutter is open, the combination of stud-and-insulating-plate is so thick (about 3 in.) that about 15% of the solar radiation that might otherwise enter the room is blocked.

Achieving good seals along all four edges of each plate may be difficult, especially if the plate becomes warped. Note that the closing torque produced by the actuator is applied to only one end of the shutter.

*Scheme 15.2a*

As above, except (a) install the actuation system near the top of the window instead of near the bottom (it is less in the way at the top), and (b) replace the pulley-and-belt system with a lever system. These improvements were developed by Levine about 1978 and are said to perform especially well.

## INDOOR TWO-PLATE SHUTTER EMPLOYING THERMOPLY (SUN SAVER SHUTTER) (Scheme 15.3)

This shutter, which makes much use of Thermoply, has been developed by C. G. Wing, J. W. Gorham, et al. of Cornerstones and/or Homesworth Corp.

The individual plate is made of two sheets of Thermoply spaced 3/4 in. apart and four straight strips of wood comprising a frame. The assembly, which is stapled and glued together, is about 1-in. thick. Each Thermoply sheet has one aluminized face and this is oriented toward the 3/4-in. airspace. Brass hinges are used. Felt weatherstripping is attached to one (or two) vertical edges of each plate. When the shutter is opened, the two plates lie close against the wall—unless the installer has elected to install the system deep within the window recess.

Sold by Homesworth Corp. of Brunswick, Maine, the equipment is delivered in the form of a kit. The homeowner cuts the material down to size, assembles it with various tools, and installs it. If desired, the faces of the panels may be covered with attractive fabric.

The shutter is sturdy. The individual hollow plate has high R-value. If the equipment is carefully assembled and accurately installed it greatly reduces heat-loss.

*Scheme 15.3a*

If the installer wishes the closed shutter to lie deep within the recess of a deeply recessed window, and yet wants the plates, when open, to lie parallel to the wall, this alternative mounting method may be used. An offsetting wooden bar is included: it is permanently attached to one side of the plate, and the hinges join it to the face of the fixed frame of the window.

A drawback is that the plate, when open and parallel to the wall, is several inches from it.

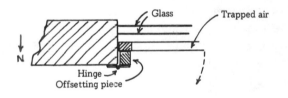

Shutter closed

Shutter open

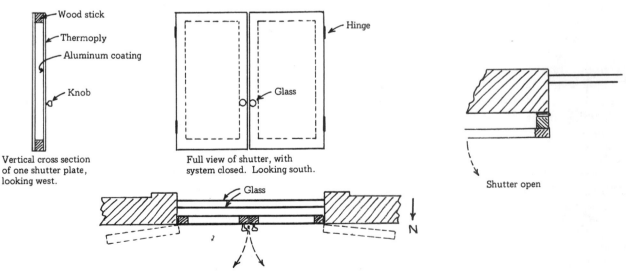

Vertical cross section of one shutter plate, looking west.

Wood stick
Thermoply
Aluminum coating
Knob

Full view of shutter, with system closed. Looking south.

Hinge
Glass

Horizontal cross section

Glass

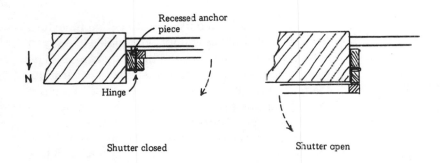

Shutter closed                    Shutter open

*Scheme 15.3b*

Here the open plates lie close against the wall. This is made possible by attaching the hinges to a recessed anchor bar instead of to the wall itself.

*Scheme 15.3c*

Cornerstones and/or Homesworth Corp. have an alternative design—bifold design—of two-plate shutter employing Thermoply. Only one plate is attached to the fixed frame of the window. The second plate is attached to the first. When such a bifold shutter is open, it takes up less wall space than a regular two-plate shutter takes up.

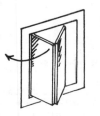

## INDOOR TWO-PLATE SHUTTER EMPLOYING RIGID FOAM CORE (Scheme 15.4)

Such a shutter is made by the British firm of Thermoblind Insulated Window Shutters Ltd. The core of each plate consists of urethane foam which is flanked by hard face sheets. Neoprene sealing strips are provided along the top and bottom edges. A positive lock is provided.

Cost: About \$4/ft², not including installation.

*Scheme 15.4a*

A somewhat similar two-plate shutter was built in 1977 by Kaiman Lee of Boston, Mass. Monsanto's insulating material called Fome-Cor was used; it includes a thin sheet of polystyrene foam. The hinges consisted of fabric-reinforced plastic adhesive strips.

## INDOOR SHUTTER EMPLOYING TWO PAIRS OF HINGED FOLDING PLATES FOR DEEPLY RECESSED WINDOW (Scheme 15.5)

This type of bi-fold shutter is found in many old houses that have deeply recessed windows, i.e., wide jambs.

There are two pairs of folding plates. When in position against the window, they cover it entirely. Buttons may be used to keep them pressed against the window.

The plates are made of wood about 1-in. thick.

To open: Turn buttons, freeing plates. Swing

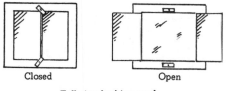

Closed                    Open

Full view looking south

Shutter closed  horizontal cross section

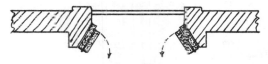

Shutter open, horizontal cross section

plates outward. Fold each pair and swing it against the window jamb.

*Scheme 15.5a*

Provide "jamb pockets" that house (conceal) the folded plates. This scheme is used in the historic Sarah Orne Jewett House in South Berwick, Maine.

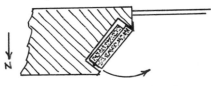

Bifold shutter in jamb pocket

*Scheme 15.5b*

Use two sets of *doubly* folded (tri-fold) plates, and use jamb-pockets. This scheme was used in Jefferson's house Monticello in Charlottesville, Virginia.

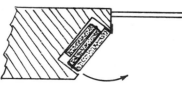

Trifold shutter in jamb pocket

*Scheme 15.5c*

As above, except that each assembly includes more plates. Thermoblind Insulated Window Shutters,

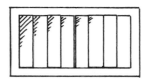

Full view, looking south.
Shutter closed.

Shutter closed, horizontal cross section     ↓ N

Shutter partly open, horizontal cross section

Shutter fully open, horizontal cross section

Ltd., of Great Britain makes shutters that have four or more plates per assembly. The plates of a given assembly are joined to each other by hinges. The main component of each plate is isocyanurate foam; various thicknesses are available. To close the shutter, the resident unfolds the assemblies so that they meet at the center of the window.

Along the base of the window there is a blade-type seal, i.e., a flexible strip mounted on edge. When the shutter assemblies are unfolded and pressed toward the window, they press against this strip.

## INDOOR SHUTTER EMPLOYING TWO PAIRS OF HINGED FOLDING PLATES FOR WINDOW THAT IS ONLY SLIGHTLY RECESSED (Scheme 15.6)

There are two folding pairs of 1½-inch-thick plates. A single extra-wide button at the middle of the windowsill is used to secure both pairs in closed position.

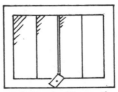

Full view, looking south

Shutter closed, horizontal cross section

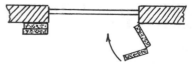

Shutter open, horizontal cross section

*Scheme 15.6a*

As above, but use three buttons.

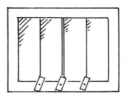

### Scheme 15.6b

Use *Thermo-Shutters* developed by J. Kachadorian et al. of Green Mountain Homes, Inc. of Royalton, Vt. Each of the four plates includes 1 in. of urethane foam sandwiched between plywood sheets. The overall thickness is about 2 in. The frame is of pine wood. The shutter is held closed by an ordinary sash lock.

### Scheme 15.6c

Use Insul Shutter developed by George Eriksen of Insul Shutter, Inc., of Silt, Colorado. Each of the four plates includes a 3/4-inch-thick, aluminum-foil-faced urethane plate. The front and back faces are of 1/8- or 1/4-in. ash plywood and there are 1/8-in. airspaces betwen the plywood and the foil-faced urethane. The perimeter consists of a frame of bass wood strips 2-in. wide and 1¼- or 1½-in. thick. Cost: About $5 or $6 per square foot, F.O.B., not including installation.

### Scheme 15.6d

Use shutters sold by Thermoblind Insulated Window Shutters, Ltd., of Great Britain. The cores of the panels include 1/4 to 1 in. of polystyrene foam.

# CHAPTER 16

# INDOOR OPAQUE SHUTTERS HINGED AT TOP

- How to Insure That a Shutter That Is Attached by Hinges at the Top Will Close Tightly Automatically

- Indoor Single Lightweight Plate Hinged at Top and Held Open by Hook Near Ceiling (Scheme 16.1)

- Indoor Single Plate Hinged at Top and Controlled by Cord and Pulleys (Scheme 16.2)

- Indoor Shutter Employing Single Plate Hinged at Top and a Counterweight That Serves Also to Provide Sealing Torque (Scheme 16.3)

- Indoor Shutter Employing Simple, Thin-Walled Cup, Hinged at Top (Scheme 16.4)

- Indoor Two-Plate Folding Shutter That Swings Up Out of the Way (Scheme 16.5)

- Indoor Accordion-Folded Set of Plates That Can Be Raised by Means of One Vertical Cord That Passes Through the Plates (Scheme 16.6)

- Indoor Shutter Employing Many Plates Each Affixed to Glazing by Means of Silicone Rubber Hinge (Silli Shutter) (Scheme 16.7)

Here I describe indoor opaque shutters that consist of one or more rigid plates, at least one of which is permanently attached by hinges at the top. Such a shutter may be simple, inexpensive, and durable.

Some useful materials for such opaque shutters are:

Thermax

Styrofoam

urethane foam

wood

plywood

Thermoply

If a plate of weak material is used, its edges may be strengthened by a wooden frame, or aluminum channels, strong tape, or a thick coat of hard, tough paint. The faces may be protected by sheets of plywood, Thermoply, or the equivalent.

## HOW TO INSURE THAT A SHUTTER THAT IS ATTACHED BY HINGES AT THE TOP WILL CLOSE TIGHTLY AUTOMATICALLY

Consider an indoor shutter that is attached to a south window by means of hinges attached in the simplest way: with the hinge pins lying in the plane of the north face of the shutter. When this shutter is released and gravity makes it swing down, it does not close tight. The pull of gravity causes it to repose at a slant. Room air can circulate between the shutter and the glass.

There are four ways of avoiding this. They are explained in detail on pages 68 and 69. In brief, they involve:

relocating the hinges so that the hinge pins lie in the plane of the shutter-face that is toward the glass,

or employing spring-biased hinges,

or employing a cantilevered weight that always exerts a positive closing torque,

or employing magnets.

Some of these strategies may be useful in some of the schemes discussed in the following sections.

## INDOOR SINGLE LIGHTWEIGHT PLATE HINGED AT TOP AND HELD OPEN BY HOOK NEAR CEILING (Scheme 16.1)

The shutter consists of a 1-inch-thick plate of Thermax or an equivalent. It is attached at the top by hinges mounted as indicated on p. 69. When open, it is held by a hook that hangs just below the ceiling. When closed, it is secured by buttons.

The shutter is so light that if it falls it will not hurt anyone. No elaborate precaution against its falling (no counterweight, for example) is required.

### Limitations

A short person cannot reach the hook.

If the shutter slips from the hook and falls, it may startle someone (but would not cause injury).

The Thermax plate may become damaged within a year or two.

*Scheme 16.1a*

As above but suspend a 2-ft length of string from the hook. A short person, wishing to lower the shutter, can pull this string northward, which will cause the hook to release.

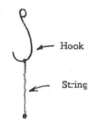

*Scheme 16.1b*

As above but dispense with the string and use, instead, a self-releasing hook. Giving the shutter a quick upward push knocks the hook away. (Proposed by J. C. Gray.)

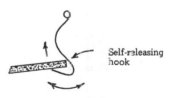

## INDOOR SINGLE PLATE HINGED AT TOP AND CONTROLLED BY CORD AND PULLEYS (Scheme 16.2)

This scheme is much like one described on p. 12 of the book *The Fuel Savers* (Ref. S-60) by Scully, Prowler, and Anderson. It has a number of drawbacks (see a later paragraph).

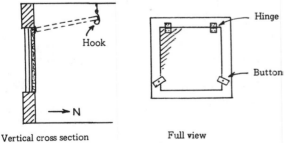

Vertical cross section looking west          Full view

The shutter consists of a 2-inch-thick rectangular plate of Styrofoam, with a 2-inch-wide wooden frame that protects all of the edges. It is hinged at the top and is controlled by a cord that is attached to the free end of the shutter. The cord runs up and over two pulleys (north pulley and south pulley) that are attached to the ceiling. The free end of the cord consists of a loop that hangs down near the window.

To open: One pulls on the cord, causing the free end of the shutter to rise northward and upward toward the ceiling. The free end of the cord (the loop) is attached to a peg at the side of the window.

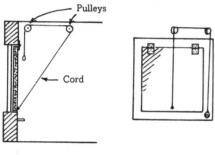

Vertical cross section
looking west    Full view

Shutter closed

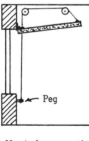

Vertical cross section
looking west

Shutter open

## Limitations

When the string is released, the shutter may descend quickly and may strike and hurt someone.

When the shutter has been lowered, there is nothing to keep it close to the window; for example, the lower edge might remain 1 or 2 inches from the window. This could be especially true if the plate is slightly warped, or if the hinges are sticky or have been mounted improperly. Such a gap would permit room air to circulate between the shutter and the glazing.

Attaching pulleys to the ceiling may be difficult.

Hauling up the shutter takes some strength.

*Scheme 16.2a*

As above, except provide, at the bottom of the window, two buttons that may be used to keep the shutter tightly closed.

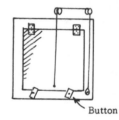

Button

*Scheme 16.2b*

As above, except (1) affix the cord to the midpoint of the west side of the shutter (and be sure that the shutter is sufficiently warp-resistant so that this asymmetric lifting scheme will not seriously distort the shutter), (2) use only one pulley, and (3) place this pulley only half as far from the south wall (and be sure that the shutter is light enough so that the inferior mechanical advantage of the haul-up system will not require one to exert a very strong pulling force).

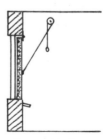
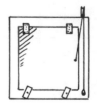

Shutter closed    Note off-to-one side
attachment point of cord,
and use of single pulley

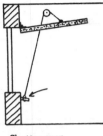

Shutter open

These changes make the haul-up system simpler, cheaper, and less conspicuous. But if the shutter is subject to warping, or is very heavy, the scheme may be unsuccessful. Also, various bad features listed above still apply: the shutter may descend very quickly and hurt someone, and hauling up the shutter may take much strength.

## INDOOR SHUTTER EMPLOYING SINGLE PLATE HINGED AT TOP AND A COUNTERWEIGHT THAT SERVES ALSO TO PROVIDE SEALING TORQUE (Scheme 16.3)

The shutter consists of a 2-inch-thick rectangular plate of Styrofoam with a 2-inch-wide wooden frame that protects the edges. The plate is hinged at the top.

A counterweight is used. It travels up and down along a vertical path near the west edge of the shutter. It is attached to a cord that is served by two pulleys. The other end of the cord is permanently attached to the middle of the shutter.

The cord is just long enough so that when the free end of the shutter has been raised almost to the ceiling, the counterweight rests on the floor.

When the shutter has been lowered (closed), one grasps the counterweight and affixes it (by a loop at the top) to a torque arm—a kind of curved hook that is attached to the middle of the shutter and projects 5 in. north from it. Thus (1) the cord becomes slack, and there is nothing tending to open the shutter, and (2) the pull of gravity on the counterweight produces a large torque on the torque arm and this tends to keep the shutter firmly closed, i.e., tightly sealed at the bottom.

To open: Disengage the counterweight from the torque arm. Pull the counterweight down.

### Comment

The system has high R-value. The seals are tight—and automatic. No buttons or latches are needed. The system is safe: when the shutter descends, it descends deliberately, thanks to the counterweight, and is not likely to hurt anyone standing in its path. The torque arm cannot hurt anyone because (1) it has no projecting end and (2) when the shutter is open, the torque arm is near the ceiling and is out of sight and out of reach.

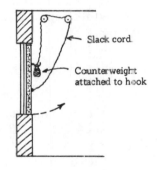

← Slack cord

Counterweight attached to hook

Scheme 16.3a

As above, except use two symmetrically-situated counterweights instead of just one. Then all of the pulleys are at the sides; none is centrally located and none is oriented obliquely.

## INDOOR SHUTTER EMPLOYING SIMPLE, THIN-WALLED CUP, HINGED AT TOP (Scheme 16.4)

In 1977 I learned from S. C. Baer the idea of insulating a window by means of a cup, or tray, made of thin metal. One possible embodiment of this scheme is described below.

Take a rectangular sheet of 0.030-in. aluminum and bend over the four edges so as to form sides about 3/4-in. high, i.e., so as to form a 3/4-inch-deep tray. Affix this by the upper edge to the indoor side of a large pane of glass. Affix it with

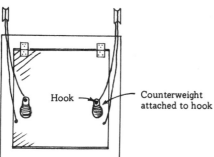

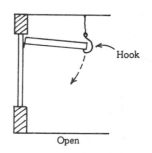

silicone cement, which provides a bond that is tough and durable but also is flexible: it can act as a hinge.

A crude handle of tape may be attached to the lower portion of the tray and a hook (hung from the ceiling) may be installed at an appropriate location to hold the tray up (open).

## Comment

If the tray is well made, i.e., if the four edges lie in a plane, the edges will automatically make a fairly good seal with the (planar) pane of glass. Insulation is provided by the 3/4-inch-thick layer of trapped air and also by the 1/4 in. of still air on the outer (north) side of the tray. The aluminum face of the tray performs the valuable service of reflecting far-IR radiation from the room.

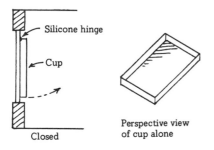

## Limitation

If the cup becomes warped, the edge seals may be very poor.

*Scheme 16.4a*

As above, except use aluminum broiling pans, pie plates, or pizza plates such as are sold in most hardware stores. Install several on each window. Attach them with transparent tape.

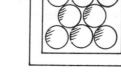

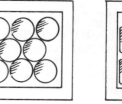

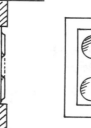

Aluminum pizza plates on window          Aluminum broiling pans on window

Here, too, a large reduction in heat-loss is provided. Crude tests made by me in January 1979 showed that when 90% of the area of a double-glazed window was covered with rectangular aluminum broiling pans, each 13 in. × 9 in. × 1.2 in., heat-loss was reduced about 50%.

## Comment

Such broiling pans cost about 50¢ each. To cover an average-size window costs only about $6. Because the pans are rectangular, they can be installed in compact array. Installation takes only a few minutes. If gaps are left between the pans, some daylight may enter during the day and thus it may be feasible to leave the pans in place night and day. If only half the glazed area is covered by pans, the heat-saving is only half as great (still considerable). The pans are non-flammable. When they are removed, they can be nested together compactly for storage.

## INDOOR TWO-PLATE FOLDING SHUTTER THAT SWINGS UP OUT OF THE WAY (Scheme 16.5)

The two plates are flexibly attached to one another by hinges. The upper plate is attached to the upper part of the window frame by hinges. (Note that there are two sets of hinges.) There is a small handle attached to the upper plate and in the lower

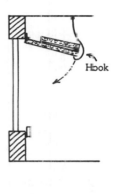

Shutter open

plate there is a correspondingly situated recess or cutout.

When the plates are in place close against the window, they are held there by buttons at the sides and by ears at the bottom. When the shutter is folded and raised, it is held up by a hook attached to the ceiling.

To open: Turn the buttons to release the shutter. Grasp the handle and pull the plates away from the window. Swing the lower plate almost 180° so that it rests on the upper plate. Swing the assembly up until it engages the hook.

## Comment

Because the shutter (when open) is folded, it does not project far out into the room.

*Scheme 16.5a*

Instead of holding up the folded shutter by means of a hook, use large arm-like spring clips (one at each side).

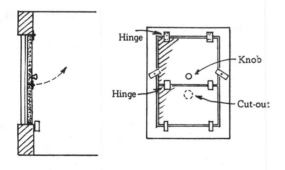

Vertical cross section
looking west, shutter closed         Full view

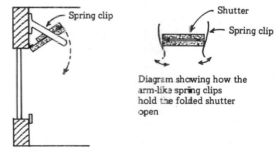

Shutter open

Diagram showing how the
arm-like spring clips
hold the folded shutter
open

To close the shutter, pull down on it, thus forcing the spring clips to release it. To open the shutter, fold it and push it strongly upward.

*Scheme 16.5b*

Use a counterweight, controlled by a cord and a pulley, to hold the shutter open. Allow the smaller (lower) plate to dangle. To close the shutter, grasp the lower plate and pull down; press the shutter against window and turn the buttons to keep the shutter tightly closed.

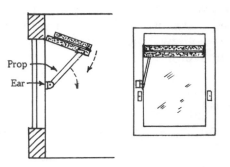

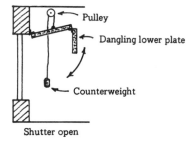

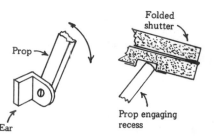

*Scheme 16.5c*

Instead of supporting the folded pair of shutters from above, support it below by means of a prop, such as a 1/2-inch-thick stick of wood. One end of the stick is permanently attached to an ear at one side of the window, and is free to swivel about the attachment point. The other end may be engaged by a sheet metal recess box taped to the under side of the adjacent portion of the upper plate.

To open the shutter, pull it north and up, fold the lower plate on top of the upper plate, raise the pair of plates, then raise the free end of the prop and engage the recess box.

**Comment**

This scheme is very simple to design, install, and operate. But its use may be limited to shutters that are so light that (1) use of an asymmetric support is permissible and (2) if the shutter should fall it cannot hurt anyone.

## INDOOR ACCORDION-FOLDED SET OF PLATES THAT CAN BE RAISED BY MEANS OF ONE VERTICAL CORD THAT PASSES THROUGH THE PLATES (Scheme 16.6)

This scheme is based largely on design made by David Wright in 1974.

Several rectangular plates of rigid foam are employed. They are attached to a long sheet of canvas, which serves (1) as a face for each plate and (2) as a flexible connection between adjacent plates. There is a hole with a smooth liner (tube) through the center of each plate. The uppermost plate is hinged to the top of the window frame. A near-vertical cord runs through the set of liners and is firmly anchored to the lowest plate.

When the set has been lowered, the entire window recess is covered (or filled) with the set of plates.

To open: Turn the crank (of the winch), thus hauling up the cord that runs vertically through the set of plates.

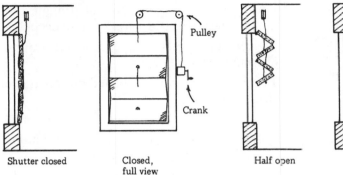

| Shutter closed | Closed, full view | Half open | Open |

## Limitation

Seals at the edges and bottom may not be tight enough.

*Scheme 16.6a*

As above, except provide vertical channels at the sides, and equip each plate edge that is always close to the window with pins that will engage the channels.

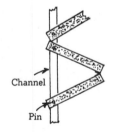

Channel

Pin

## INDOOR SHUTTER EMPLOYING MANY PLATES EACH AFFIXED TO GLAZING BY MEANS OF SILICONE RUBBER HINGE (SILLI SHUTTER) (Scheme 16.7)

This shutter, developed by S. C. Baer et al. of Zomeworks Corp. in 1977 and further worked on by Michael Sherson (according to *Alternative Sources of Energy* No. 27, Aug. 1977, p. 24) is simple, cheap, and effective but leaves the windows encumbered at all times.

The shutter consists of a set of plates which define air pockets at the north side of the glazing.

Each plate consists of a thin (0.030 in.) strip of aluminum about 7-in. wide and as long as the glazing is wide. The main axis is horizontal, but the strip lies in a vertical plane. Each edge is bent at about 45° from the body of the strip. Each end of the strip is closed with a wooden block "end filler."

The upper (bent) edge is affixed to the window glazing by means of adhesive (e.g., clear silicone rubber sealant, or #760 silicone rubber sealant), which forms an improvised hinge.

When the strip is bent upward (about 40 or 50 degrees, and with the upper edge still firmly attached to the glazing), direct solar radiation can enter the room freely.

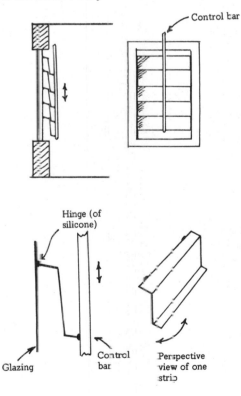

Control bar

Hinge (of silicone)

Glazing

Control bar

Perspective view of one strip

When the strip is allowed to descend, the lower edge touches (and overlaps 3/8 in. of) the upper edge of the next-lower strip; thus the glazing area is completely covered by the set of strips. The set of strips defines a set of 1-inch-thick air-pockets.

The strips are operated by means of a long, slender bar that is attached to the lower edge of each strip. Operation is somewhat like that of a venetian blind.

R-value is said to be about 2 or 3.

Cost of materials:  About $1.20/ft$^2$.

## Limitations

If any aluminum strip is attached to the glazing at a slightly wrong angle or position, the seal with respect to a neighboring strip may be poor; a gap may remain.

If any edges of the strip are not straight, or if the strip becomes warped, gaps will occur.

The strips encumber the window at all times, obstructing view.

The window cannot be opened (unless it is of casement type?)

Will the informal hinges fail after a few years?

The system is not applicable if the window area is broken up by many muntins.

*Scheme 16.7a*

Fill the "pocket" with a long strip of styrofoam attached to the aluminum strip.

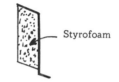

Styrofoam

# CHAPTER 17

# INDOOR OPAQUE ONE-SHEET ROLL-UP SHADES AND ROLL-UP SHUTTERS

- Indoor Standard Roll-Up Shade (Scheme 17.1)
- Indoor Standard Roll-Up Shade With Filler Sticks (Scheme 17.2)
- Indoor Standard Roll-Up Shade With Vertical Clamping Bars Controlled at Top and Bottom (Scheme 17.3)
- Indoor Standard Roll-Up Shade With Vertical Clamping Bars Controlled at Center (Scheme 17.4)
- Indoor Extra-Wide Roll-Up Shade That Has Edge Seals of Informal Type (Scheme 17.5)
- Indoor Extra-Wide Roll-Up Shade With Vertical Clamping Bars (Scheme 17.6)
- Indoor Roll-Up Shade With Two Vertical Channels and, at Top, a Horizontal Sealing Lip (Windoseal by Lynn Economic Opportunity, Inc.) (Scheme 17.7)
- Indoor Roll-Up Shade Employing Thin Quilt and Two Rollers (ATC Window Quilt) (Scheme 17.8)
- Indoor Roll-Up Shade Employing Thin Quilt, Two Rollers, and Wooden Channels (Helio Construction Shade) (Scheme 17.9)
- Indoor Roll-Up Shade Sealed Along Edges by Tiny Magnets (Zomeworks Roll-Away Magnetic Curtain) (Scheme 17.10)
- Indoor Roll-Up Shutter Employing Rigid Slats (Thermo-Shade) (Scheme 17.11)

Here discussion is concentrated on indoor, opaque, one-sheet shades that consist mainly of flexible material and can be rolled up. I deal first with the most common indoor window covering in current use. Then some radically new kinds are described.

Much attention is given to seals at the vertical edges and the bottom of such shades. Tight seals can greatly increase the heat-saving. Many kinds of seals have been developed and used successfully, and many as-yet-untried kinds appear promising. Some comparisons of seal types are presented in Chapter 21.

If this chapter is long, it is because the subject of well-sealed roll-up shades is an important one. Such shades can produce large heat-savings, yet cost relatively little. Also they are attractive, durable, safe, easy to install, and solve the storage (parking) problem in an ideal manner.

Note: Roll-type devices that include many rigid strips are called shutters and usually are situated outside, not inside, the building. (See Chapter 10.)

## INDOOR STANDARD ROLL-UP SHADE (Scheme 17.1)

The commonest shade in use today is a simple roll-up shade that employs a plastic (usually vinyl) film or a sheet of fabric and is white, yellow, or green in color. The roller, of wood or cardboard, is about 1 in. in diameter. At one end of the roller there is a tiny cylindrical axle that turns with the roller. At the other end there is a tiny rectangular-cross-section strip that does not turn; when the roller turns (and the strip does not turn) a helical spring within the roller is wound up or unwound. It is wound up when the shade is pulled down (closed). When thus wound up, it produces enough torque so that it can roll up (raise) the shade. (It cannot roll up the shade until a tiny pawl is released—by a downward jerk on the shade).

If such a shade is pulled far down (so that the lower edge is 1/2 in. from the sill), and if the shade is wide enough so that the gaps at either side are only about 1/2 in., the amount of room air that circulates into the space between glass and shade is small and the heat-loss through the (double-glazed) window is reduced by about 7 to 25%, according to 20 tests I made on the east window of my den in evenings in February 1979. Results varied greatly from one evening to the next. Typical value: 20%.

### Comment

The closed shade is, on the average, fairly far from the glazing. It may be 2½ in. from the upper sash glazing and 1½ in. from the lower sash glazing. Thus no face seal is achieved. At best, there is an edge seal, but this is likely to be poor because of the big gaps at the sides, bottom, and top.

Warning: If one pulls the shade down farther, until the lower end rests on the sill, the upper portion of the shade becomes slack and may belly out, greatly widening the gaps at the sides. Ideally, the shade is pulled down just far enough so that it barely touches the sill.

*Scheme 17.1a*

As above, except reverse the roller: turn it end-for-end. Then the shade, when pulled down, lies much

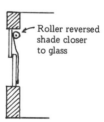

Roller reversed shade closer to glass

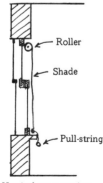

Vertical cross section shade closed

Roller
Shade
Pull-string

Full view, with no shade in use

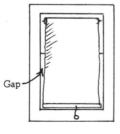

Gap

Shade closed

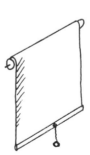

Perspective view of shade

Pawl
Spring
Axle

X-ray view of roller

closer to the sashes—and very close indeed to the horizontal frame member at mid-height. Thus circulation of air here is somewhat discouraged.

## INDOOR STANDARD ROLL-UP SHADE WITH FILLER STICKS (Scheme 17.2)

This scheme is like the previous one except that a major effort is made to provide good seals at the vertical edges of the shade. Two vertical wooden *filler sticks*, attached to the window jambs, are used. The function of such sticks is to (1) provide a single, smooth surface for the edge of the shade to press against, and (2) arrange for this surface to be slightly off-vertical so that the weight of the shade automatically causes the shade to press (if only very slightly) against this exposed surface.

Each stick extends from the sill almost to the top of the window. The stick is tapered, being about 1/2-in. wide at the top and about 2-in. wide at the bottom, as indicated in the accompanying sketch. The back of the stick is cut away (relieved) to accommodate the step from upper sash to lower sash, and the cut-away region is made extra long so as to permit the lower sash to be raised a few inches (should the resident wish to open the window a little). If the gap between the edge of the shade and the jamb is 1/2 in., the stick should be about 1- or 1½-in. thick, so that the shade-edge will overlap the stick by 1/2 in. or more. If the bottom of the stick is flush with the edge of the sill, and if the shade is pulled down so that its lower edge bears against the sill, a fairly good seal is achieved here.

If the shade proper (or the stiffener stick incorporated in its lower edge) is heavy, the side seals are fairly tight.

If the window is of single-sash type, no cutouts are needed in the filler sticks.

### Scheme 17.2a

As above but add vertical tension cords to improve the edge seals. Obtain two 6-foot-long nylon cords and two 2-lb weights, one to be attached to each cord. Attach the free end of each cord to a tiny horizontal east-west post (e.g., a long nail or long screw) affixed to the window jamb at such a location that the cord lies close against the pertinent filler stick. Then when the shade is closed, the two cords press the shade edges firmly against the filler sticks, thus providing excellent seals here. It is helpful if the north faces of the filler sticks are slightly curved—slightly crowned.

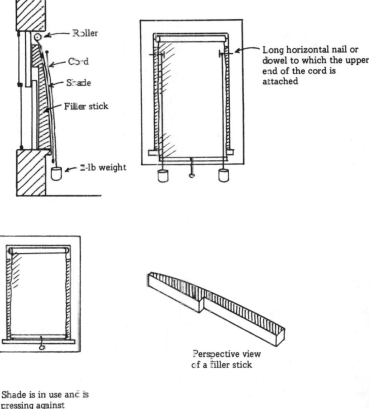

Roller

Cord

Shade

Filler stick

2-lb weight

Long horizontal nail or dowel to which the upper end of the cord is attached

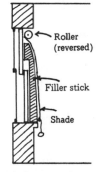

Roller (reversed)

Filler stick

Shade

vertical cross section looking west, shade closed

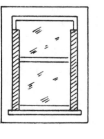

Full view, looking south. No shade in use.

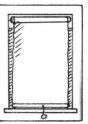

Shade is in use and is pressing against filler sticks and sill

Perspective view of a filler stick

Note: The weights are so light, and the forces involved so slight, that the shade can be raised or lowered in the usual way, almost as if the weights and cords were not there.

*Scheme 17.2b*

As above, except replace the vinyl sheet with a shiny aluminum sheet, such as Foylon, Dura-Shade, Astrolon, or aluminized mylar. These materials are described in Chapter 5 and Appendix 5.

## Comment

The combination of an aluminized shade and good seals at sides and bottom should produce a large saving of heat, if the window in question is double glazed and reasonably tight. The saving may be 40 to 60%. When the shade is aluminized, the use of good seals at the edges may be highly cost-effective. It does not matter if one side of the aluminized sheet is painted or coated with fabric; the important requirement is that at least one face—either face—be naked (or near- naked) aluminum, i.e., be immediately in contact with air only (at least 1/2 in. of air). How many years will an exposed aluminum surface remain highly reflective to far-IR? I do not know. Many years, I guess. Even if reflectivity with respect to visual-range radiation falls off considerably, the reflectivity with respect to far-IR may remain high. (Why not use just a plain sheet of aluminum foil? Because it may soon crumple and tear, especially if it is repeatedly rolled up to a diameter of only 1 inch. Until it becomes damaged, such a sheet can do a superb job of cutting heat-loss, according to tests made long ago by others and tests made recently by me.)

*Scheme 17.2c*

As above, except use a sheet of special material that is transparent to solar radiation (including visual-range radiation) but that strongly reflects far-IR radiation. A shade of such material could be left down day and night: during the day it would (1) permit the occupants of the room to see out, (2) admit much solar radiation, and (3) discourage heat-loss by radiation, conduction, and convection. At night it would greatly reduce heat-loss.

## Comment

Such material is under development (see Appendix 7), but it is not yet readily available. When it becomes available, will its optical properties be close to ideal, or will they fall far short? Will the cost be acceptable?

When such material is available and is used in roll-type shades, the use of good edge seals will pay off handsomely.

Because such material is transparent to visual-range radiation, it does not provide visual privacy. A supplementary (opaque) shade may be needed.

Of course, when such material becomes available many persons may put it to use as permanent glazing—to be left untouched day and night, and perhaps summer and winter also. Use on roll-up shades may not be the best use.

## INDOOR STANDARD ROLL-UP SHADE WITH VERTICAL CLAMPING BARS CONTROLLED AT TOP AND BOTTOM (Scheme 17.3)

In this scheme, applicable to double-sash windows and to an ordinary roll-up shade mounted in the window recess, short filler sticks are employed and sealing pressure is applied by vertical straight bars (clamping bars). (If the scheme is to be used with a single-sash window, no filler sticks are needed.)

The filler sticks, which serve just the upper sash and are attached to its side members, are just thick enough to provide surfaces that are flush with the lower sash. Thus the filler sticks simplify the sealing problem by providing a single vertical plane for the shade edges to lie against and for the vertical clamping bars to press against. Each filler stick is attached by means of a single screw near the top; thus each is free to be swung laterally and upward to permit the lower sash to be raised a foot or so when and if one wishes to open the window. Alternatively the stick may be attached to the sash by means of a 1/4-inch-diameter dowel that projects (in slightly downward direction) from the stick and engages a corresponding hole (say, 3/8 in. in diameter and 3/8-in deep) in the sash.

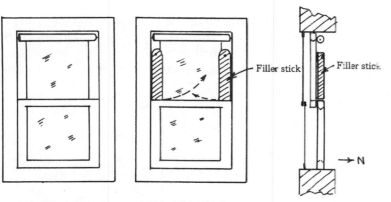

Typical window

Filler stick installed
in such a way that they
can be swung upward to
allow lower sash to be
raised

Vertical cross
section of window
and filler stick

Each clamping bar is a long, rectangular-cross-section strip of wood (or equivalent) attached to a window jamb. The attachment means is such that the clamping bar can either (1) press firmly (under the influence of gravity) south and downward so as to press the shade against the sash and filler stick, thus providing a positive seal, or (2) remain 1/2 in. away from the sash and filler stick, i.e., remain disengaged so as to allow ample room for raising or lowering the shade.

In designing the controls at the upper and lower ends of the bar, I have given much attention to the compatability of the controls, i.e., to avoid having them "fight each other" if one is the least bit out of synchronism with the other. From extensive experimentation I found that such "fighting" is almost sure to occur if both of the controls are rigid (or precise, or positive). The solution? Employ one control that is rigid and one that is compliant over a large range.

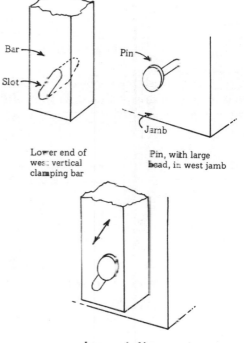

Lower end of
west vertical
clamping bar

Pin, with large
head, in west jamb

Lower end of bar secured
to jamb by a pin in slot

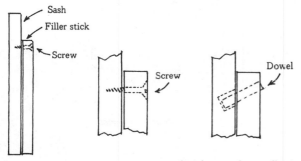

Attachment by a single screw

Attachment schemes allowing
instant removal

headed pin that projects horizontally in east or west direction from the window jamb and engages a diagonal slot in the bar. When the bar is free to descend, the pressure of the pin against the slot face pushes the bar toward the shade and the filler stick. When the bar is raised, the pin forces the bar away from the shade and filler stick.

The upper end of the clamping bar is held captive by a rigid fixed horizontal arm attached to the face of the fixed frame of the window. The arm permits the bar to move up or down and allows it 1/2-in. motion toward or away from the sash. There is a vertical leaf spring attached to the upper end of the bar, and when the bar moves down, this spring is compressed by the arm and makes the bar press against the edge of the shade. The arm serves an additional purpose; it holds the bar in up position when the bar is raised an inch or two and is pulled away from the window so that the lower edge of the leaf spring catches on the arm.

To close shade: Pull the shade down until the lower edge rests on the sill. Push the tops of the clamping bars south to free them from the arms. *To open shade:* Pull the clamping bars upward and north until the upper ends of the bars are engaged by the arms; then snap up the shade in the usual way.

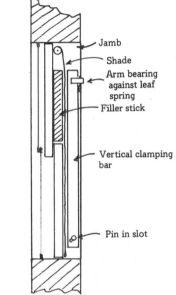

Cross section of window, shade, etc., showing vertical clamping bar pressing against shade to form edge seal

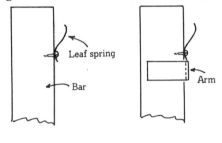

Upper end of west vertical clamping bar

Bar held up by arm

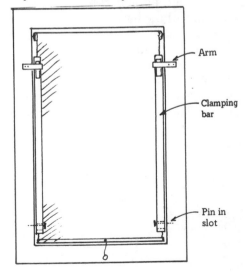

Bar pushed toward window by spring and arm

This end of arm is attached to fixed frame os window

Full view of shade and vertical bars loosely attached to jambs

*Scheme 17.3a*

Here the vertical clamping bars are supported and guided in a somewhat different manner. Each bar is linked to the window jamb by two pivoting strips of galvanized iron. When the bar is in operating (clamping) position, the forces exerted on it by gravity and by the pivoting strips urge it toward the window. When the bar is moved upward and away from the window, it leaves a 1/2-in. space in which the shade is free to travel up or down. The pivoting strips are free to turn at least 60 degrees and thus can swing past the uppermost (dead center) position and permit the bar to move northward and rest against a fixed ear. The holes in the lower pivoting strip are somewhat oversize; thus this strip is less positive (more compliant) than the other one.

To open shade: Move each bar upward and away from the window until the bar rests against the ear. Then snap the shade up. *To close shade:* Pull the shade down and push the bars toward the window so that they tend to fall and push against the edges of the shade.

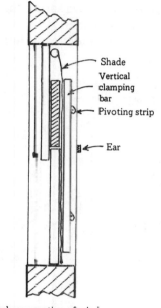

Vertical cross section of window equipped with vertical clamping bars attached by pivoting strips

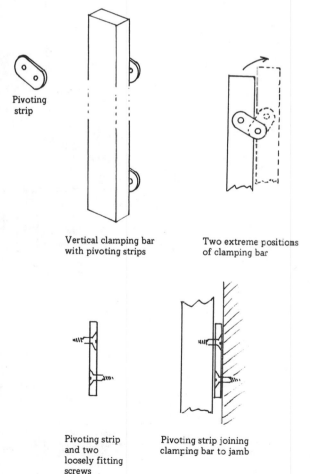

Pivoting strip

Vertical clamping bar with pivoting strips

Two extreme positions of clamping bar

Pivoting strip and two loosely fitting screws

Pivoting strip joining clamping bar to jamb

*Scheme 17.3b*

Here each vertical clamping bar can be rotated about a vertical axis defined by the top and bottom pivots. The top pivot is held by a fixed block attached to the fixed frame of the window; the bottom pivot engages a hole in the sill. The bar has a cam-like cross section, and when the bar is rotated 90 degrees, by means of a handle, it presses against the edge of the shade. Sealing torque is provided by rubber bands; one end of each band is attached to an off-axis hook at the end of the bar and the other end of the band is attached to the fixed frame of the window.

To open shade: Turn each handle 90 degrees *away* from the center of the window, to free the shade. Then snap the shade up. *To close shade:* Pull the shade down and turn each handle 90 degrees toward the center of the window, squeezing the edge of the shade between the clamping bar and the window sash.

## Comment

The clamping operation is especially forceful and firm, and it has the additional feature of urging the edges of the shade away from the vertical center-line of the window, i.e., of flattening the shade (overcoming any tendency to bulge in horizontal cross section).

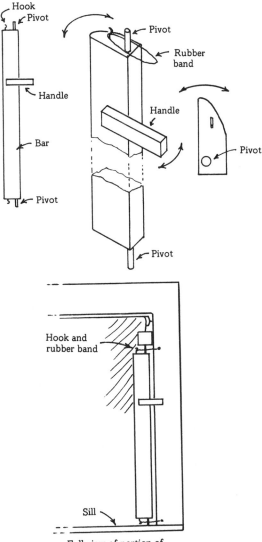

Full view of portion of
window and a pivoting
clamping bar

*Scheme 17.3c*

Use clamping bars that are supported from the side by leaf springs attached to the sides of the fixed frame of the window. When relaxed, the springs hold the clamping bars 1/2 in. from the sashes. Near the center of each clamping bar there is a small ramp which may be urged toward the window by rotation of a button that is attached to the face of the fixed frame of the window and bears against the ramp.

The weight of each bar is supported by the upper leaf spring, which is especially wide and is attached to the fixed frame of the window by several screws.

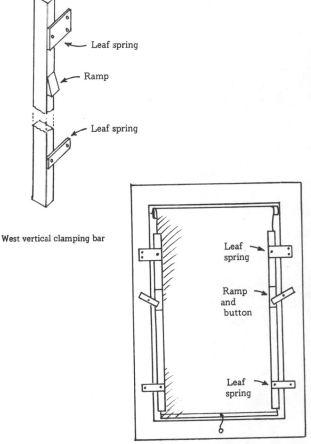

West vertical clamping bar

Full view looking south
Pressure of buttons against ramps
insures tight edge seals

To open shade: Disengage the buttons (thus allowing the bars to move 1/2 in. away from the window), then snap up the shade. *To close shade:* Pull the shade down and turn and tighten the buttons (thus forcing the bars toward the window).

## INDOOR STANDARD ROLL-UP SHADE WITH VERTICAL CLAMPING BARS CONTROLLED AT CENTER (Scheme 17.4)

This scheme is designed for windows that are only 3- or 4-ft tall. Loosely attached to each jamb is a vertical wooden clamping bar that is confined only by a single screw that passes through a special bracket near the center of the bar and is firmly fixed in the window jamb.

The bar has two positions of stability: (1) a position where it hangs freely from the screw and makes little or no contact with the shade (so that the shade can be raised or lowered without resistance), and (2) a position where it is wedged against the screw and presses against the shade, sealing it against the sash, or, if there are upper and lower sashes, sealing it against the lower sash and an upper filler stick.

To change the position of the bar from (1) to (2), one merely gives the bar a southward tap. To make the change from (2) to (1) one grasps the bracket and pulls it north and upward.

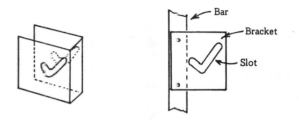

### Comment

If the window were very tall this scheme might perform poorly: the bar might bend enough so that the seal would at some places be poor. Also the (free) upper and lower ends might move laterally (east and west) excessively. Use of a scheme providing controls at both ends of the bar is preferable if the window is very tall.

### Scheme 17.4a

As above, except use a very different design of bracket. Use a piece of sheet metal that has been bent around so as to embrace the bar. The two main faces of the bracket are appropriately slotted.

This bracket is stronger and more easily attached. Also, the two faces of the bracket serve in place of washers.

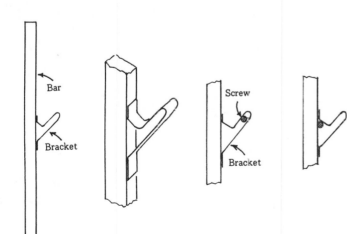

| Bar and bracket | Perspective view of bar and bracket | The two positions of bar and bracket relative to screw |

Full view showing how each bar is attached to jamb by a single screw and two washers

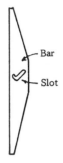

*Scheme 17.4b*

Make the central region of the bar extra wide and incorporate the slot in it. Then no bracket is needed.

## INDOOR EXTRA-WIDE ROLL-UP SHADE THAT HAS EDGE SEALS OF INFORMAL TYPE (Scheme 17.5)

Here the shade, of ordinary vinyl material, is extra wide and is mounted in front of the fixed frame of the window rather than within the window recess. Being about 6 inches wider than a recessed shade, it overlaps the side members of the fixed frame by several inches. Because it overlaps them and is so close to them, a rough approximation to an edge seal is achieved here. The roller is mounted just above the window recess, close to the top frame member. The bottom of the shade just touches (or almost touches) the sill.

Heat saving is about 15 to 30%, according to a few crude measurements made by me.

### Limitations

Mounting the roller in this position requires special brackets.

The shade may roll up crooked, damaging the edges of the shade. (When a shade is mounted within a window recess, crooked roll-up is largely prevented.)

Appearance may be unsatisfactory.

There are no *tight* seals.

Heat-loss by far-IR radiation is scarcely reduced at all.

*Scheme 17.5a*

As above, but use filler sticks also. Each is attached to the fixed frame of the window, i.e., beside the window, not within the recess. Each filler stick is tapered, being thicker at the bottom; thus the north face is slightly off-vertical and accordingly gravity causes the shade to make contact with this face along its entire length, thus providing a rough seal. The bottom of the filler stick extends out to the

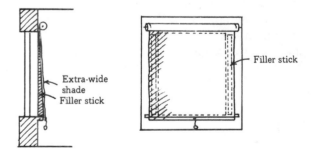

Extra-wide shade
Filler stick

Filler stick

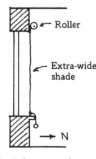

Roller

Extra-wide shade

→ N

Vertical cross section looking west

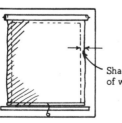

Shade overlaps fixed frame of window about 2 in.

Full view , looking south

edge of the sill, and the bottom of the shade is pulled down slightly below the sill and makes contact with it.

*Comment* No cutouts in the filler sticks are needed. Even if the window is of double-sash, double-hung type, the simplest shape of filler stick suffices. The sticks do not interfere with the opening or closing of either sash of the window.

*Scheme 17.5b*

As above, except use filler sticks that have convex north faces and use weighted cords to press the shade against these sticks. The upper end of each cord is held by a small post that is based off-side so as not to interfere with the shade. The weighted cords interfere only slightly with raising or lowering the shade.

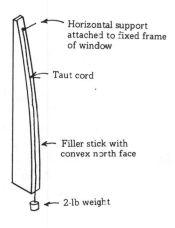

*Scheme 17.5c*

As above, except dispense with the added weights and, instead, install a 1-lb horizontal bar in the bottom edge of the shade. (This will require using a stronger wind-up spring in the roller.)

*Scheme 17.5d*

Replace the simple shade with an aluminized shade. This will reflect far-IR radiation and thus will considerably increase the heat-saving. Rough tests made by me in March 1979 indicated that a

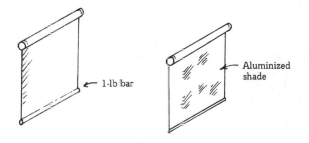

single sheet of aluminum foil, well sealed at the edges and flanked only by air, can reduce heat-loss through a double-glazed window by 65%.

## INDOOR EXTRA-WIDE ROLL-UP SHADE WITH VERTICAL CLAMPING BARS (Scheme 17.6)

Here use is made of vertical clamping bars situated beside the window rather than in the window recess. The bars press the edges of the extra-wide shade against the face of the fixed frame of the window (or against the wall adjacent to the window). Again the bars can be in either of two positions: south position, sealing the edges of the shade, or upward-to-north position leaving the shade free to be raised or lowered. The sealing force is provided by gravity. The control at one end of each clamping bar is positive and at the other end is more compliant, so that the two controls do not "fight."

Attached to the top of each bar is a strip, or loop, of steel, which includes a 60-degree-sloping segment and a notch. The strip is engaged by a broad-headed pin attached to a fixed wooden block off to the side. When the sloping segment of the steel strip bears down on the pin, the pin's counterforce urges the strip toward the south (toward the wall); if the bar weighs a few pounds, the southward force is also of the order of a few pounds. It is this force that insures that the edge of the shade is well sealed. When the bar has been raised so that the pin engages the notch, the bar remains about 1/2 in. from the wall.

The lower end of the bar, which extends below the lower edge of the shade, is notched so as to form, on the side toward the wall, a kind of ramp.

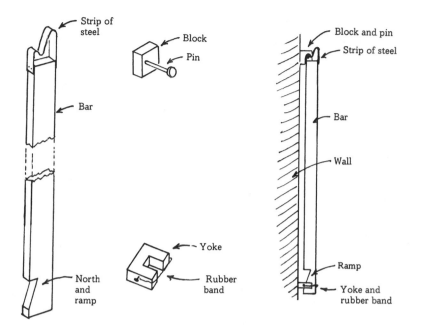

Views of a vertical clamping bar and its supports

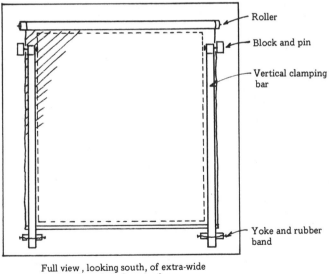

Full view, looking south, of extra-wide shade and vertical clamping bars

The ramp bears against the base of a transverse yoke attached to the wall and is urged toward the ramp by a rubber band attached to the yoke. When the bar is raised, the lower end of the bar is forced away from the wall because of the motion of the ramp relative to the yoke. When the bar is lowered, its lower end is free to move toward the wall and the rubber band forces it to do so.

To open shade: Grasp the bars, raise them and engage the notches—so that the bars remain 1/2 in. from wall. Snap up the shade. *To close shade:* Pull it down, then disengage the bars so that they may fall downward and southward toward the wall.

## Comment

Note that the single act of raising a bar an inch or two causes both ends to move away from the wall, releasing the shade, and the single act of lowering the bar forces both ends toward the wall and seals the edge of the shade. Any tendency for the bar to rotate about a vertical axis is resisted by the top and bottom supports. The upper support prevents upward and downward overtravel. The force exerted by the rubber band at the bottom of the bar is small relative to the (gravitational) force effective at the top; thus if the adjustment, or synchronization, of the upper and lower mounts is incorrect, no harm is done: the force at the top dominates and insures overall correct operation. When the bar is lowered, its position is determined solely by the wall; thus slight maladjustments of the mounting devices do not affect the tightness of the seal. It is important that the bars and the wall be flat.

Obviously, instead of using a rectangular cross section wooden bar, one could use a bar of steel or aluminum and the cross section could be round (use a pipe, say), rectangular, I-shaped, L-shaped, or U-shaped. If a very lightweight bar is used, the downward force of gravity could be supplemented with a downward-pulling spring.

### Scheme 17.6a

As above, but use a different kind of mounting for the upper ends of the vertical bars: for each bar use a rigid bent steel strip that "flies over" the roller and, above it, engages a curved rod that has a sloping segment and also a notch. When the sloping segment is engaged, the rod is urged toward the wall. When the notch is engaged, the rod hangs 1/2 in. from the wall.

### Scheme 17.6b

As above, except attach the upper end of the vertical bar by means of a crank that turns about a horizontal east-west axis defined by a horizontal hole in a block of wood attached to the wall. If the moving end of the crank (the end that engages the bar) is at its highest point and is then moved south (toward the wall), it begins to descend and, while descending, strikes the wall and presses against it. If it strikes the wall when the crank angle is 60 or 70 degrees from the vertical, the force against the wall is about twice the downward pull of gravity. That is, there is a helpful gain factor, as regards force. If, contrariwise, the crank is at the top of its travel and is then moved away from the wall, it can turn through 180 degrees, and the vertical bar then hangs well clear of the wall.

Obviously, a steel bracket could be used in place of the wooden block.

This type of edge seal was proposed by Douglas Barr in January 1979.

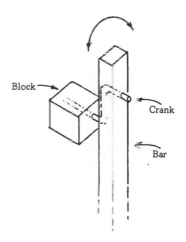

overall correct operation. *(labels: Block, Crank, Bar)*

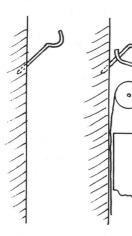

Hook on wall          Bar engaged

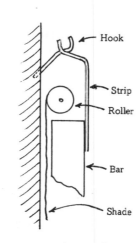

*(labels: Hook, Strip, Roller, Bar, Shade)*

Bar disengaged

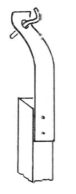

Perspective view

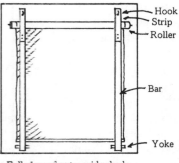

Full view of extra-wide shade and fly-over vertical clamping bars

*(labels: Hook, Strip, Roller, Bar, Yoke)*

*Scheme 17.6c*

As above, except use a different kind of control at the lower ends of the bars. For each bar provide a weak leaf spring that is attached to the wall area below the bar and, when the bar is in lowered position, urge the tapered lower end of the bar toward the wall. When the bar is raised, the spring loses contact with the bar, allowing it to hang free of the wall. At the upper end of the spring there are flanges that at all times restrain the lower end of the bar from moving sidewise, i.e., east or west.

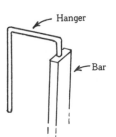

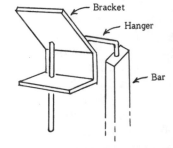

Hanger and upper end of vertical clamping bar

Upper support assembly

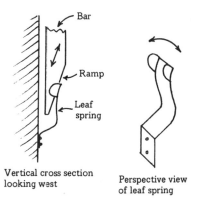

Vertical cross section looking west

Perspective view of leaf spring

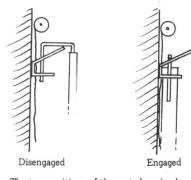

Disengaged          Engaged

The two positions of the east clamping bar

*Scheme 17.6d*

Support the upper end of each vertical clamping bar by means of a 1/4-inch-diameter steel rod that has been bent into an inverted-U shape. One end of the rod (hanger rod) is attached to the upper end of the bar; it engages a vertical hole in the bar and is glued in place. The other end is captive within a bracket that is attached to the wall. The bracket has a special shape: its upper member is ramp-like. When the bar is swung away from the wall (away from the edge of the shade), it is perforce raised, gaining potential energy. When fully raised, its hanger rod rests on the horizontal edge of the top of the ramp and thus is in equilibrium. To clamp the edge of the shade, one merely rotates the bar until the hanger rod rests on the steeply sloping ramp, whereupon the bar continues to rotate and descend until it presses firmly against the edge of the shade, forming a positive seal.

The lower end of the bar is held by a somewhat similar hanger rod and bracket, except that this rod is of more flexible material so that even if the upper and lower systems are not perfectly matched no harmful rivalry will arise.

*Scheme 17.6e*

This scheme is much like one developed about 1978 by C. G. Wing. Each of the vertical clamping bars is supported at each end by a fixed vertical pivot, or axle, and can rotate at least 45 degrees. At one extreme of rotation, one edge of the bar presses against the edge of the shade and the wall, producing a seal. At the other extreme, the bar is 1/2 in. from the wall. The rotational position is controlled by a leaf spring, one end of which is attached to the wall; the other edge presses northward (away from the wall) against a cam attached to the vertical bar. The shape and position of the cam are such that (1) when the bar is approximately in contact with the wall, a strong torque is exerted by the spring, to insure tight seal, and (2) when the bar is turned 45 degrees, no torque is exerted. A small stop is included to prevent the bar from being turned more than 60 degrees.

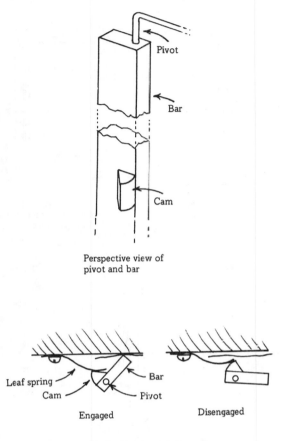

Perspective view of
pivot and bar

Engaged          Disengaged

Horizontal cross sections of
bars and leaf springs

To release the bar and free the shade: grasp the rod and turn it 45 degrees. To reinstate the seal, turn the bar at least 20 or 30 degrees the other way, then release it, allowing the spring to complete the closing.

### Scheme 17.6f

As above, except use hinges that include strong springs. The hinges support the vertical clamping bars and press them toward the wall, to seal the intervening shade. Such hinges are used in the Roman Shade II developed by the Center for Community Technology. No pivots at the top and bottom of each bar are needed, and no leaf springs are needed.

### Scheme 17.6g

Here, each vertical clamping bar is supported from the side by thin plates of galvanized iron that are mounted in such a way (or bent in such a way) that, when no special force is applied, the bar is 1/2 in. from the wall. To apply a force urging the bar toward the wall, one grasps the knob on the free end of a small helical spring (the other end of which is permanently attached to the wall), draws it over the pertinent support plate, and engages an adjacent slotted post. (One could use a rubber band: merely pull on it and loop one end around the post—or around a simple hook.)

### Comment

The support plates govern the lateral and vertical positions of the bars in positive manner, but they are highly compliant as regards to-and-fro (north-south) position; the springs control this.

### Scheme 17.6h

As above, except use somewhat thinner, more compliant support plates and provide latches cap-

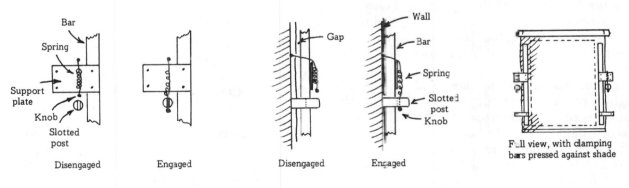

Disengaged          Engaged          Disengaged          Engaged          Full view, with clamping bars pressed against shade

Views of the main support and pressure system

able of keeping the plates pressed strongly toward the wall. Further details are indicated in the sketches.

To make the seal: Press hard on the support plates so as to bow them sufficiently so that they will be captured by the latches and will keep the bars pressed toward the wall. To release the bars: press the latch tabs upward.

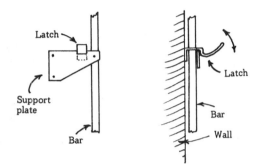

Latch holding clamping bar close against wall

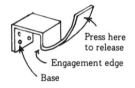

Perspective view of latch

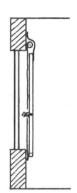

Vertical cross section
looking west

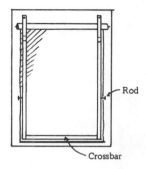

Full view, looking south.
Clamping bar assembly pressing
against three edges of
shade

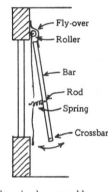

Clamping bar assembly
swung away from window
to release shade edges

*Scheme 17.6i*

Join the lower ends of the two vertical clamping bars with a horizontal crossbar. Equip the upper ends of the clamping bars with extensions, or fly-overs, that pass in front of the roller and are attached to the wall in hinging manner, or merely by means of steel strips that are flexible. The mid-regions of the bars are urged toward the wall by helical springs attached to outrigger rods that consist of 2-inch-long round-head screws.

To open shade: Grasp the crossbar and pull it away from the wall a few inches, thus releasing the edges of the shade, and snap shade up. *To close shade:* Again pull the crossbar away from the wall, pull the shade down, and release the crossbar.

## INDOOR ROLL-UP SHADE WITH TWO VERTICAL CHANNELS AND, AT TOP, A HORIZONTAL SEALING LIP (WINDOSEAL BY LYNN ECONOMIC OPPORTUNITY, INC.) (Scheme 17.7)

This device, called *Windoseal,* was developed in 1977 by Technology Development Corp. of Boston, Mass. Later it was promoted and marketed by Lynn Economic Opportunity, Inc., of Lynn, Mass.; however the program was later discontinued.

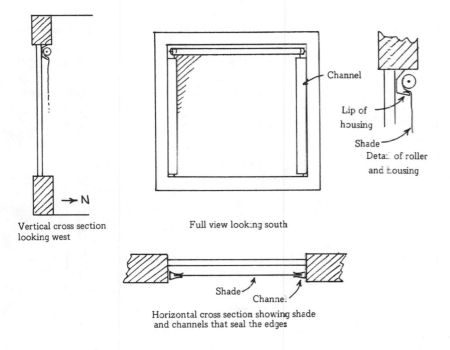

Vertical cross section
looking west

Full view looking south

Channel

Lip of
housing

Shade

Detail of roller
and housing

Horizontal cross section showing shade
and channels that seal the edges

Shade    Channel

The heart of the device is a 0.004-inch-thick sheet of Melanex, an aluminum-coated plastic sheet that has a transmittance of about 28%. The 1-inch-diameter roller, at the top of the window, includes a rewind spring and pawl. Attached to the window jambs are two vertical, 1-inch-deep channel strips of flexible PVC. Each is very slender; the sides closely embrace the faces of the Melanex sheet, thus providing a fairly tight seal. A strip magnet included in the lower edge of the shade attracts a corresponding metal piece integral with the window sill, making a seal there.

A slender housing partly surrounds the roller. The housing makes contact with the uppermost part of the fixed frame of the window, and a lip along the lower edge of the housing forms a seal with the upper part of the shade.

Cost for a typical window: About $18, not including installation.

## INDOOR ROLL-UP SHADE EMPLOYING THIN QUILT AND TWO ROLLERS (ATC WINDOW QUILT) (Scheme 17.8)

This shade, made by Appropriate Technology Corp. and called *Window Quilt*, is attached to the north face of the fixed frame of the window; that is, it is adjacent to, but not in, the window recess.

It includes five layers, joined every few inches by ultrasonic fabricweld. The central layer, 0.001-in. aluminized mylar, is flanked by 3-oz, 3/16-inch-thick, polyester fiberfill which in turn is flanked by dacron cloth. Overall thickness: about 3/8 in. Each of the vertical edges consists of an integral bead strip 1/4 in. in diameter. The assembly is dry-cleanable.

Affixed to one end of the 1¼-inch-diameter main roller is a 4-inch-diameter grooved plastic disk (pulley) on which a control cord is wound. There is no wind-up spring or pawl. To raise the shade (roll it up), one pulls on the cord, then locks it in a jamb roller provided. The diameter of the rolled-up quilt is about 5 or 6 inches.

A second roller (idler roller, also 1½ in. in diameter and of wood) is situated adjacent to the main roller and very close to the top of the fixed frame of the window. The quilt passes between the idler roller and the window frame.

As the shade is lowered (by releasing the cord and allowing the shade to be pulled downward by gravity), the right and left beads of the shade slide vertically within tough plastic channels that are

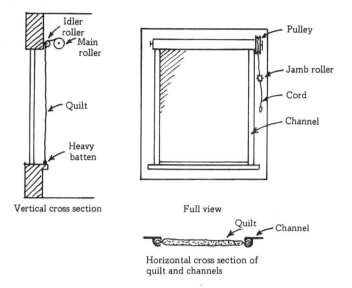

Vertical cross section

Full view

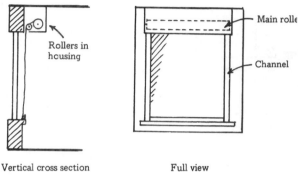

Horizontal cross section of quilt and channels

cemented, nailed, or stapled to the wide members of the fixed frame of the window. The body of the channel is 1/4-in. wide inside and the gap between the channel lips is about 1/8 in.

The bottom edge of the quilt consists of an integral stiffener stick, or batten, which is heavy and is provided (on the bottom) with compressible weatherstripping. This rests on the window sill.

Cost: About $3.75/ft$^2$, not including installation.

An excellent description of this device appears in the October, 1979, Popular Science Monthly, p. 98.

### INDOOR ROLL-UP SHADE EMPLOYING THIN QUILT, TWO ROLLERS, AND WOODEN CHANNELS (HELIO CONSTRUCTION SHADE) (Scheme 17.9)

Plans and instructions for this shade are provided by Helio Construction of 190 E. 7 St., Arcata, CA 95521. Cost of plans in 1979: $3. The homeowner himself must procure the materials needed.

The quilt (e.g., of fabric and polyester batting) is rolled up on a roller that is about 6 in. in diameter. An idler roller keeps the shade close to the plane of the window frame. Both rollers are enclosed in a large wooden housing attached just above the window. The large roller is controlled by

a rope wound around a pulley integral with the roller.

When the shade is closed its sides and stiffened lower edge are confined within channels made of wooden strips. The side channels may be concealed beneath pleated curtains.

To open: Pull rope.

Vertical cross section

Full view

Horizontal cross section of quilt and channel

### INDOOR ROLL-UP SHADE SEALED ALONG EDGES BY TINY MAGNETS (ZOMEWORKS ROLL-AWAY MAGNETIC CURTAIN) (Scheme 17.10)

This scheme, developed by Zomeworks Corp. in 1978, has been described in the New Mexico Solar Energy Association's *Southwest Bulletin* of July 16, 1978, p. 16. The heart of the invention is a scheme for sealing the right and left edges of a roll-up shade by means of slender linear arrays of tiny magnets.

The individual array of magnets is 1/2-in. wide, 1/16-in. thick, and many feet long. The magnetic material is made in strip form by 3M Co. and is somewhat flexible. To greatly increase the flexibility, so that the array can be rolled up with a diameter as small as one inch, the strip is cemented to a rubbery backing strip and the strip proper is then cut through at 1/3 in. intervals.

Within each of the 1/3-inch-long pieces of magnet the pattern of north and south poles is ingeniously arranged so that as the strip is rolled

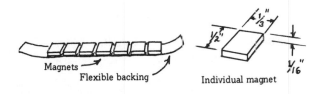

Magnets    Flexible backing    Individual magnet

up upon itself, the individual pieces—instead of repelling each other and resisting compact roll-up—attract each other and facilitate roll-up.

Each long, slender magnetic array, attached to an edge of the shade, is attracted by a strip of steel along one side of the window—either steel that is already part of a metallic window frame or an added strip of light-gauge steel that is cemented to a wooden or plastic frame.

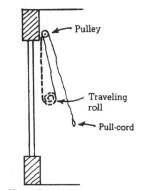

Vertical cross section

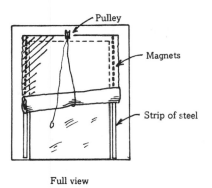

Full view

To open:  Pull on pull cord, thus rolling up the shade and the strips of magnets. A gentle pull suffices because there is no sliding motion and the tiny magnets are pulled away from the steel strips sequentially, not all at once.

## INDOOR ROLL-UP SHUTTER EMPLOYING RIGID SLATS (THERMO-SHADE) (Scheme 17.11)

Made by Solar Energy Components, Inc., and called Thermo-Shade, this shutter is similar to certain roll-up, slat-type shutters described in Chapter 10. It may be installed in the window recess or in front of it.

The heart of the shutter is a flexibly interlocked set of hollow plastic (PVC) slats, each 1½-in. wide and 1/2-in. thick at the crown. Each end is closed by a cap which carries two tubular prongs 5/16 in. in diameter and 3/8-in. long. The prongs slide up and down in channels permanently attached by neoprene foam adhesive tape to the fixed frame of the window. The shade may be rolled up onto a 4-inch-diameter horizontal roller within a housing at the top of the window. Roll-up is facili-

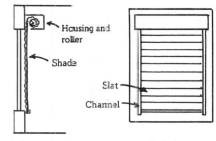

Vertical cross section          Full view

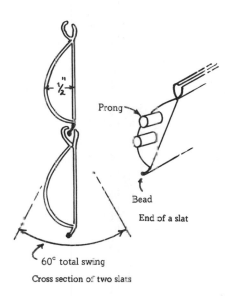

60° total swing

Cross section of two slats

tated by a coil spring within the roller. When the shade is down, a bottom seal is formed by a strip of foam rubber. The shade may be opened to any extent desired, and is held at the desired position by brake shoes at each edge; the shoes may be released by squeezing the tabs of a central control knob.

An optional motor drive, controlled by photocells, is available.

Cost: About $5/ft² F.O.B., including shutter proper, channels, roller, etc., but not including motor drive or installation.

*Scheme 17.11a*

Use somewhat similar system developed by Alan Ross of Brattleboro Design Group and described in *Solar Age*, Aug. 1979, p. 19. The slats, 3/4-in. thick and 1½-in. wide, consist mainly of Thermax. The upper and lower edges of the slats are of wood, and are V-shaped. The slats are wrapped with jute twine. The shutter is controlled by a crank and nylon cord. The design is being modified with a view to reducing cost.

*Other Alternatives*

One could presumably make use of some of the outdoor-type roll-up shutters described in Chapter 10.

# CHAPTER 18

# INDOOR OPAQUE MULTI-SHEET ROLL-UP SHADES

- Indoor System Employing Two Shades on One Recessed Roller (Scheme 18.1)

- Indoor System Employing Two Shades on One Extra-Wide (Non-Recessed) Roller (Scheme 18.2)

- Indoor System Employing Two Shades on Two Recessed Rollers (Scheme 18.3)

- Indoor System Employing Two Simple Shades on a Recessed Roller and a Non-Recessed Roller (Scheme 18.4)

- Indoor System Employing Two Simple Shades on Two Non-Recessed Rollers (Scheme 18.5)

- Indoor, Three-Sheet, Three-Roller Shade System (Insealshaid Made by Ark-Tic-Seal Systems Inc.) (Scheme 18.6)

- Indoor Five-Sheet Shade (Made by Insulating Shade Co., Inc.) (Scheme 18.7)

- Indoor, Four-Sheet, Roll-Up, Self-Inflating Shade (Insulating Curtain Wall) (Scheme 18.8)

- Indoor, Two-Sheet, Two-Roller Shade System That Can Constitute a Duct That Facilitates Solar Heating (Scheme 18.9)

An easy way to reduce heat-loss at windows with little fuss and bother is to use two roll-up shades instead of one. Adding a second shade is easy, and the added saving of heat is moderately large or large depending on the materials used and the tightness of seal. If one of the shades is aluminized and edge seals are used, the heat-saving may be about 40 to 60% as compared to about 10 or 20% provided by an ordinary, unsealed roll-up shade.

Addition of a second roll-up shade is especially cost effective if edge-sealing equipment has been added for the existing roll-up shade and could equally well serve two shades.

The variety of ways of adding a second shade is enormous. A score or so of ways are described in the following pages. I deal first with schemes requiring only one roller; then I discuss two-roller schemes. I describe schemes in which both shades—or just one, or neither—are situated within the window recess.

## INDOOR SYSTEM EMPLOYING TWO SHADES ON ONE RECESSED ROLLER (Scheme 18.1)

Remove the existing roller and shade, lay it on the floor, and fully unroll it. Then attach another shade to the roller with staples, tacks, or adhesive tape. The second shade may be of the same material as the first, or of different material; if the first is of thick, absorbing material, the second can be of less substantial material; whether it is thick and absorbing or thin and transparent is of little consequence, presumably. But if the first is thin and transparent, the added shade should be thick and/or absorbing. To avoid overloading the return spring within the roller, one may favor adding a shade that is lightweight. If the material is very thin, flexible, and compliant, both sheets may roll up together easily and unroll easily.

If each shade, when unrolled, tends to be somewhat wavy, so that the two sheets hang about 1/2 in. apart, this is all to the good: more regions of trapped air exist, and the vertical circulation of air in these regions is impeded.

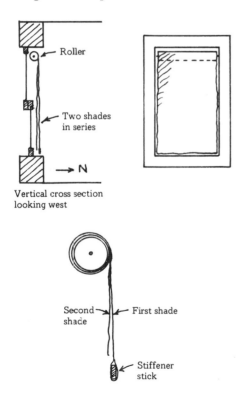

Vertical cross section looking west

Second shade — First shade

Stiffener stick

Attention should be given to the appearance of the added shade, its flammability, and its durability.

Adding a horizontal weighting bar to the lower edge of the second shade may be helpful. A single pull cord (attached to the first shade) may suffice.

### Scheme 18.1a

As the added shade, use an *aluminized* sheet—an aluminized plastic film or aluminized fabric. E.g., use aluminized mylar, Astrolon, or Foylon. Because such a sheet has high reflectance for far-IR radiation, the heat-saving it provides is large. (I believe it makes little difference which of the shades is aluminized, or which face of a given shade is aluminized. But it is highly desirable that there be an airspace at least 1/2-in. thick immediately adjacent to the aluminum surface.)

### Scheme 18.1b

Install a fixed, horizontal 1/2-inch-diameter rod parallel to the roller and a few inches below it. Use it as a separator bar: when the pair of shades is unrolled each passes on a different side of the bar, with the result that there is a 1/2-in. layer of trapped air between the shades. This further increases the heat saving.

The bar is so high up that it continues to separate the shades even when they are nearly fully rolled up. (Instead of a bar, one could use a piece of thick wire or rope.)

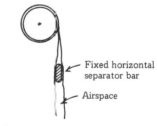

Fixed horizontal separator bar

Airspace

### Scheme 18.1c

Turn the roller end-for-end so that the shades, when unrolled, lie about 1 in. closer to the upper-sash glazing. Thus the thickness of the region of air between the shades and the glazing is smaller (for example, 1 in. rather than 2 in.) and convective flow here is reduced.

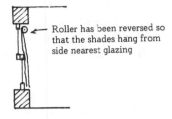

Roller has been reversed so that the shades hang from side nearest glazing

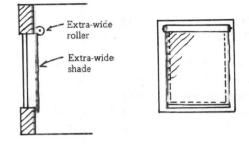

Extra-wide roller

Extra-wide shade

*Scheme 18.1d*

Make the added sheet slightly wavy, or corrugated, with the corrugations horizontal. The corrugations insure that there is an airspace between the two shades and discourage circulation of air in this space.

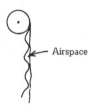

Airspace

*Scheme 18.1e*

Seal the edges of the shades (or at least the edges of the north one). Use any of the pertinent sealing systems described in Chapter 17, e.g., schemes employing filler stick or vertical clamping bars situated within the window recess and close to the jambs.

**INDOOR SYSTEM EMPLOYING TWO SHADES ON ONE EXTRA-WIDE (NON-RECESSED) ROLLER (Scheme 18.2)**

Mount the extra-wide roller very close to the wall-areas (or window casing areas) immediately to the east and west of the window. Employ, as a north shade, one that is about 4 or 6 in. wider than the window and thus will overlap these adjacent areas by a few inches. The south shade may be equally wide, or may be considerably narrower; its width is not crucial inasmuch as the edges of the north shade provide seals that, in a sense, serve both shades. If the north shade lies within 1/8 in. of the wall, the seal here is at least moderately helpful in preventing convective currents of room air from circulating between the shades or between the shades and the glazing. The lower edges of the

shades may rest on the window sill or may bear against the north edge of the sill.

*Scheme 18.2a*

As above, but use at least one shade that is aluminized.

*Scheme 18.2b*

Use a south shade that is narrow enough to fit within the window recess, and employ a horizontal separator bar that will insure that this shade lies in the recess and is separated at least 1/2 in. from the north shade.

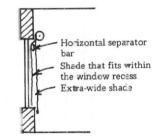

Horizontal separator bar

Shade that fits within the window recess

Extra-wide shade

*Scheme 18.2c*

Employ devices that will positively seal the vertical edges of the north shade (or both shades). Use filler sticks and/or vertical clamping bars.

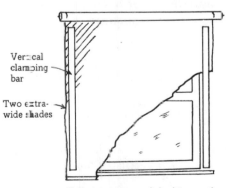

Vertical clamping bar

Two extra-wide shades

Full view (cut-away), looking south

## INDOOR SYSTEM EMPLOYING TWO SHADES ON TWO RECESSED ROLLERS (Scheme 18.3)

Here two rollers are used: one for each of the (recessed) shades. The operation is a little more time-consuming: there are two shades to adjust rather than just one. But it may be feasible to raise or lower both shades at once, with a single motion of one hand. The danger of overburdening any roller, or its built-in spring, is avoided. Also the added flexibility—the option of employing either shade alone—may be of some value.

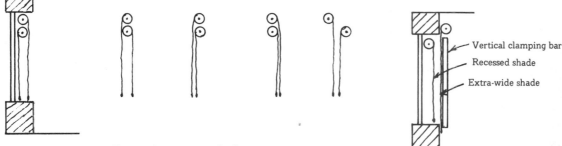

Choices of orientations of rollers

Each roller may be installed in normal or reversed orientation according to circumstances such as convenience in installing the support brackets or the thickness of the intervening airspace desired.

*Scheme 18.3a*

Employ at least one aluminized shade.

*Scheme 18.3b*

Install a horizontal bar to control the thickness of the airspace between shades.

*Scheme 18.3c*

Make at least one of the sheets wavy, to impede convective air currents.

*Scheme 18.3d*

Employ positive seals at the vertical edges. Use filler sticks and/or vertical clamping bars.

## INDOOR SYSTEM EMPLOYING TWO SIMPLE SHADES ON A RECESSED ROLLER AND A NON-RECESSED ROLLER (Scheme 18.4)

If one already has an ordinary roll-up shade, which is recessed, one may wish to buy an additional shade that is much wider and will overlap the window casing. In some crude experiments I found that such schemes, using one recessed shade-and-roller and one overlapping shade-and-roller, worked well. Operating the shades was simple, and the second shade increased the saving appreciably.

— Vertical clamping bar
— Recessed shade
— Extra-wide shade

*Scheme 18.4a*

Employ at least one aluminized sheet.

*Scheme 18.4b*

Employ positive seals at the vertical edges. Use vertical clamping bars such as are described in Chapter 17.

## INDOOR SYSTEM EMPLOYING TWO SIMPLE SHADES ON TWO NON-RECESSED ROLLERS (Scheme 18.5)

My crude experiments showed that this scheme is also successful. The operation was straightforward and, thermally, two shades performed a good deal better than one alone. However, the appearance of the window region was unsatisfactory. Presumably one could install a valance that would conceal the two rollers.

*Scheme 18.5a*

Employ at least one aluminized sheet.

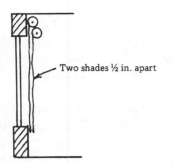

*Scheme 18.5b*

Employ positive seals at the vertical edges.

## INDOOR, THREE-SHEET, THREE-ROLLER SHADE SYSTEM (INSEALSHAID MADE BY ARK-TIC-SEAL SYSTEMS INC.) (Scheme 18.6)

This device, made by Ark-Tic-Seal Systems Inc. and called *Insealshaid*, can operate in several modes and provide several benefits in addition to insulating the window at night.

There are three shades, three rollers, multi-channel vertical tracks, a housing, and other components. All three shades are transparent. Each

consists mainly of polyester and is attached to a roller within the housing. The north shade is dark gray (60% to 70% absorbing). The intermediate shade is optically clear (but with a slight gold cast) and transmits virtually no ultraviolet radiation. The south shade includes a vacuum-deposited aluminum layer between laminated films of mylar and reflects 80% of the incident solar radiation. This shade is 0.0035-in. thick. The others are 0.005-in. thick.

Each vertical edge of each shade moves up and down within one channel of a vertical three-channel track of extruded aluminum. Each channel includes opposed strips of nylon pile that form a fairly tight seal. Additional pairs of pile strips situated close below the rollers provide seals at the top of the window. The lower edge of each shade terminates in a bottom clamping bar to the underside of which a 1/4-inch-diameter compressible plastic tube is secured to provide a seal against the common horizontal base plate. Thus all four edges of each shade are well sealed.

At the top of the system there is a housing which encloses the three rollers. The housing is designed so that when a damper in the base of the housing is open, air from the space between the

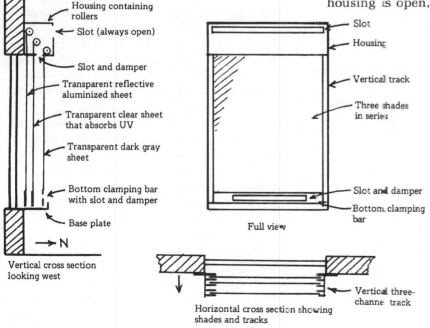

Full view

Horizontal cross section showing shades and tracks

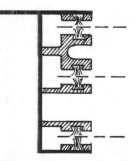

Detailed horizontal cross section of vertical three-channel track equipped with three pairs of nylon-pile strips. All sketches are simplified and not to scale.

north shade and the intermediate shade can flow upward into the housing via a long slot and can then emerge into the room via another long slot in the housing's north face. The damper, which is of extruded aluminum and is centrally pivoted, is controlled by a bimetallic strip which opens the damper when the temperature reaches about 85°F. How does room air enter the space between the north shade and the intermediate shade? Via a slot in the clamping bar at the base of the north shade. On the south side of this slot there is another bimetallic-strip-controlled damper; this opens when a specified temperature is reached, such as 70°F.

The south shade may be raised or lowered manually, with the aid of a pulley and a nylon cord, without the need to first raise the other shades.

Although most cost effective when applied to a single-glazed window, the system is applicable also to double-glazed windows. The shade system may be mounted within the window recess if this is several inches deep; otherwise it is mounted against the fixed frame of the window.

Many operating modes are available:

During cold winter nights, use all three shades and thus greatly reduce heat-loss.

During sunny days in winter,

use no shades; allow much solar radiation to penetrate deep into room, or

use the north and intermediate shades so that the air between these shades will become hot and will circulate (via top and bottom vents) to the room, or

use just the intermediate shade—to admit solar radiation but reduce conductive loss of heat.

During winter days with intermittent clouds, use the north and intermediate shades so that when the sky is cloudy, the shades and closed vents will provide a high degree of insulation, and when the sky is clear much hot air will be delivered to the room via the top and bottom vents.

During hot sunny days in summer, use the intermediate and south shades; together they exclude 80% of solar radiation and reduce conductive inflow of heat.

At all times when glare is a problem or excessive ultraviolet radiation may cause curtains, rugs, etc., to fade, use the intermediate shade, which transmits no ultraviolet radiation. Some persons may choose to use this shade at all times—night and day, summer and winter —to protect rugs, curtains, etc., and to reduce conductive flow of heat.

Cost: About $7.50 per ft² F.O.B., for a typical window. This cost includes shades, rollers, tracks, housing, dampers, etc., but not installation.

## INDOOR FIVE-SHEET SHADE (MADE BY INSULATING SHADE CO., INC.) (Scheme 18.7)

This remarkable shade, capable of very great reduction in heat-loss, was developed by Insulating Shade Co., Inc., of Branford, Conn. It was demonstrated at a 9/9/77 exhibition in Hartford, Conn., and was featured in the January 1979 issue of *Popular Science Monthly*.

Although the shade includes five sheets in series, there is but one roller, for example a 1¼-inch-diameter wooden roller with internal helical heavy-duty wind-up spring and an automatic latch (pawl).

Of the five sheets, the (front and back) outermost ones are of vinyl plastic that has an aluminum coating on the innerside. The three interior sheets consist of plastic that has been aluminized on both sides. Thus there are 8 aluminum layers in all. At the top, all five sheets are secured to the roller. At the bottom, the two exterior sheets are joined to form a kind of envelope or bag. The bottom of the envelope contains a horizontal wooden bar that provides stiffness and weight.

When the shade has been pulled down (to cover the window), spacers come into play to maintain spaces of about 1/2 in. between the sheets. Each spacer is a plastic strip about 1½-in. wide and as long as the shade is wide. In cross

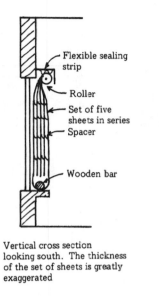

Vertical cross section looking south. The thickness of the set of sheets is greatly exaggerated

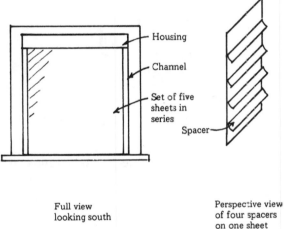

Full view looking south

Perspective view of four spacers on one sheet

section the spacer is curved, having about the same curvature as the surface of the roller. Spacers are affixed to four of the sheets, and on each of these the spacers are about 7 in. apart on centers. Besides keeping the sheets 1/2 in. apart, the spacers serve as partitions to divide the space between adjacent sheets into several subregions. The overall thickness of the set of shades is about 2 to 2½ in.

The right and left edges of the shade assembly are confined within vertical channels, or boxes, 2½ in. in inside width. The bottom of the envelope, with a compressible sealing strip attached to it, rests on the window sill. The gap above the roller is closed by means of a pair of flexible sealing strips

attached to the upper part of a canopy, or housing, that contains the roller.

When the shade is rolled up, all five sheets roll on the one roller. The spacers, having curvature matching that of the roller, occupy almost no space; thus the entire rolled-up assembly is only about 2¾ in. in diameter.

To close the shutter, one pulls down strongly on a cord attached to the center of the bottom of the envelope. To open the shutter, one gives a jerk to the cord and then pays it out at a moderate speed (about 1 ft/sec) as the spring within the roller rolls up the set of sheets automatically.

Cost: The cost of a complete system for a 5-foot-wide, 6-foot-high window, including shade proper, roller, side channels, top housing, and sealing strips for top and bottom is about $110, not including installation, i.e., about $4/ft². (?)

Cost of the shade alone: about $2.75/ft².

## Comment

The system enormously reduces radiation and conduction losses. In fact one may wonder whether, with four air spaces and eight aluminum-coated surfaces, the shade is not overdesigned. As regards convection: the edge seals are comparable to those of some of the other high-performance thermal shades; yet one may wonder whether the seals are tight enough to significantly reduce in-leak or out-leak of air through a leaky window.

## Limitations

The cost is high.

Considerable labor is involved in installing the device.

There is a possible need to widen the window sill to accommodate the very thick shade.

Some persons may not like the appearance of the side channels and top housing.

If the shade is allowed to roll-up very rapidly, air is trapped between the various sheets, causing them to billow out.

The shade (with its wooden bar at the bottom) is so heavy, and the wind-up spring so strong, that raising and lowering the shade is some-what difficult.

One may wonder whether such a relatively complicated system will last for many years without deterioration.

As of July, 1979, some of the key components (side channels) were not available.

### INDOOR, FOUR-SHEET, ROLL-UP, SELF-INFLATING SHADE (INSULATING CURTAIN WALL) (Scheme 18.8)

This unique device, invented and patented by Ronald Shore of Thermal Technology Corp., has the registered name *Insulating Curtain Wall*. Sometimes it is called *Self-Inflating Curtain*. It is intended for use with large-area windows in cold climates and is well suited for use with Trombe walls or other structures that will impart heat to one side of the curtain on winter nights. An excellent description of the device appears in *Popular Science* of October 1979, p. 99.

There are four sheets mounted on a single 3-inch-diameter roller. Each sheet is 0.004-in. thick. The outer sheets are of polyester on which aluminum has been vacuum deposited, and a layer of rip-stop nylon cloth is included. The nominal reflectance of the aluminum layer is 93%. The inner sheets, which include aluminized fabric, have a nominal reflectance of 68%. The aluminum layers are on the south side of the south sheet and the north sides of the other sheets.

At the sides of the window there are 5-inch-wide vertical channels which confine and guide the edges of the shade.

The two outermost sheets of the shade are joined at the bottom to form a kind of bag. Near the bottom of the bag there are several horizontal slots (in all four sheets) that permit air to enter or leave. Each slot is 30-in. long and 3-in. wide. Included in the bottom of the bag is a 1½-inch-diameter tube that weights the bag down and helps form a tight seal against the window sill.

The roller is turned by a small electric motor situated within one end of the roller. The enclosing cylindrical housing is 6 to 8 in. in diameter.

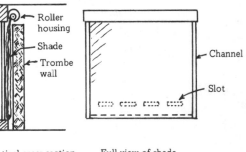

Vertical cross section of shade and Trombe wall

Full view of shade alone, looking south

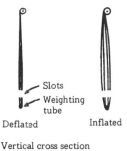

Vertical cross section shade: deflated and inflated. The thickness is exaggerated.

The shade has a unique self-inflating capability. Specifically, its thickness increases greatly, growing from a fraction of an inch to about 5 inches finally, when and if the air within it is hotter than the air adjacent to it. When the shade is inflated, there are air spaces 1/2- to 1-in. thick between successive sheets.

Why does the set of sheets expand? One might guess the reason to be that air, when heated, expands. But this is not a correct explanation, because any such expansion would amount to only a fraction of an inch. The correct explanation is one that I explained in a letter of about Dec. 15, 1977, to Ronald Shore. A chain reaction occurs when the air within the shade is hotter than the adjacent unconfined air. When the confined air is hotter, it is less dense: a 5-foot-high column of such air weighs less than a similar column just outside the shade, and accordingly the pressure along the vertical centerline of the confined column is more nearly uniform than the pressure along the vertical centerline of the unconfined (cooler, more dense) column. Thus if the pressures at the bases of the columns (near the slots) are equal, then the pressures at the tops of the columns are unequal; specifically, the pressure at the top of the confined column exceeds that at the top of the unconfined column. Consequently the air in the upper part of the confined column (and in the upper region of the shade as a whole) tends to expand, pushing the walls of the bag outward. In summary, the thickness of the bag increases—and at the same time additional air flows into the bag, via the slots near the base, to accommodate this increase in volume of the shade. The newly introduced air in turn becomes hotter, and less dense, and accordingly the tendency of the bag to expand continues. As the bag expands, more and more air enters via the slots. The process continues until the stiffness or weight of the bag, or the confining effect of the side channels, stops further expansion.

## Comment

The three air spaces defined by the expanded set of four sheets discourage heat-flow by conduction and the four aluminized layers discourage energy-flow by radiation. When the assembly expands and presses against the adjacent vertical surface or surfaces, excellent face seals exist. In summary, the thermal performance of the shade is excellent. The R-value is said to be about 5 before the shade is inflated and twice as great when inflation is complete.

The shade can be used in summer to exclude energy. The shade is lowered, and solar radiation heats it and makes it expand.

Cost: For a large window, such as one 20-ft wide and 13-ft high, the shade costs $4.50/ft$^2$, retail, F.O.B., not including side channels. For an 8-foot-wide, 6-foot-high window the cost is about $8/ft$^2$.

## Limitations

The device is not recommended for use on small windows.

There are situations in which little or no inflation occurs. But even then, thanks to the many aluminized sheets, a fairly high degree of insulation is provided.

## INDOOR, TWO-SHEET, TWO-ROLLER SHADE SYSTEM THAT CAN CONSTITUTE A DUCT THAT FACILITATES SOLAR HEATING (Scheme 18.9)

If, adjacent to the double- (or single-) glazed window, there are two roll-up shades, the designer can arrange for the two shades, when unrolled, to form the two main sides of a vertical duct. If at least one of the shades is dark colored, it may absorb much of the solar energy passing through the window, and the resulting heat may be transported, by natural or forced flow of air in the duct, to the room as a whole or to a thermal storage system (say in the attic or basement). Dampers, operated manually or automatically may be employed to direct the hot air from the duct to the room or to the storage system.

If the shades are rolled up, the incoming solar energy penetrates deep into the room, heating it promptly.

At night the rolled-down shades reduce heat-loss through the window.

In summer the rolled-down shades can be used to exclude solar radiation, especially if the south face of the south shade is aluminized. If it is not aluminized and the air between the shades becomes very hot, this air may be vented to the outdoors: the venting may assist the inflow, elsewhere, of cool outdoor air.

Many persons have invented schemes of this general type and there are many published articles on the subject. (See, for example, U.S. Patent 3,990,635 of 11/9/76 by J. W. Restle, A. J. Algaier, and G. R. Krueger.)

I describe no specific system in detail because (a) any such scheme is somewhat complicated and (b) the variety of systems that may be proposed is enormous. Furthermore, various between-glazings modifications may be proposed and also various modifications involving rigid plates.

# CHAPTER 19

# INDOOR DEVICES OF OTHER TYPES

- Curtains That Are Permanently Fixed and Sealed at Top and Sides and Are Opened by Diagonally Upward Pulls (Scheme Described by Scully, Prowler, and Anderson) (Scheme 19.1)

- Curtain-and-Liner That Can Be Slid Horizontally (Scheme Employing Liner Sold by Wind-N-Sun Shield, Inc.) (Scheme 19.2)

- Curtains That Are Permanently Affixed and Sealed to Sides of Window and Can Be Slid Along a Horizontal Bar at Top and Then Sealed at Center and Bottom by Means of Magnets (Scheme Developed by Conservation Concepts Ltd.) (Scheme 19.3)

- Curtains That Are Permanently Affixed and Sealed at Sides of Window Frame and Can Be Slid Along a Horizontal Rod at Top and Can Be Sealed at Bottom by a Long Heavy Pipe (Scheme 19.4)

- Indoor Shade Held Up by Hooks at the Top (Scheme 19.5)

- Indoor Multi-Cloth-Sheet That Can Be Collapsed Downward (Scheme 19.6)

- Indoor Quilt That Can Be Accordion Folded (Sun Quilt) (Scheme 19.7)

- Indoor Honeycomb-Type Shade That Is Guided by Vertical Splines and Can Be Hoisted (Scheme Developed by J. J. Anderson, Jr.) (Scheme 19.8)

- Indoor Device Employing Many Slender Vanes (Scheme 19.9)

- Indoor Device Employing Big Thick Vanes (Skylid) (Scheme 19.10)

- Indoor Shutter Employing Two Plates, One of Which Is Hinged at the Bottom (Scheme 19.11)

- Indoor Shutter Employing a Plate That Is Hinged at the Bottom and, When Swung Downward, Forms a Worktable (Scheme 19.12)

Here I deal with all other kinds of indoor shutters and shades applicable to vertical windows. Most of these devices involve heavy curtains, or quilts. Some employ small permanent magnets or Velcro. Some require special vanes.

● Warning concerning curtains that lack bottom seals: J. F. Peck of the University of Arizona has warned (P-130) that full-window-height curtains that are not sealed at the bottom may perform poorly. A person standing beside a curtained window may have the impression that curtains make the room warmer: he feels warmer. However, when he places his hand in the gap between the bottom of the curtain and the window or wall, if he feels a stream of cold air, he should question whether the curtains are as effective as they appear.

Note concerning magnets: Magnets as thin as 1/16 in. may be very effective; being thin, they cost little. Some retain their magnetism for decades. Some types include much plastic and are flexible; others have an adhesive layer on one face. There is no need to use magnets in pairs; it is more economical to use a pair consisting of one magnet and one piece of ordinary steel.

Note concerning Velcro: Instead of using magnets to join the edge of a shutter to a window frame, one can use Velcro, a pair of tapes one of which includes thousands of tiny, flexible, hook-like plastic fibers and one which includes thousands of tiny plastic loops. When the two tapes are pressed together (lightly or strongly—it does not matter), the set of hooks engages (randomly, statistically) the set of loops. Thus the two tapes are found to be fairly strongly bonded together, face to face. A strong pull can separate them again. Such operation can be repeated thousands of times; the hooks and loops do not deteriorate.

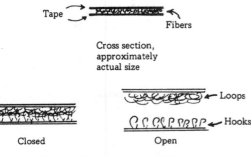

Cross section, approximately actual size

Closed          Open

Cross section, greatly magnified

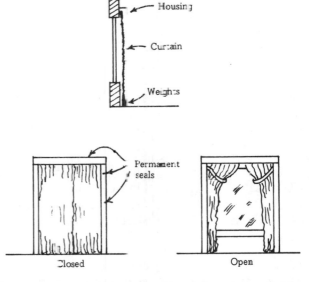

## CURTAINS THAT ARE PERMANENTLY FIXED AND SEALED AT TOP AND SIDES AND ARE OPENED BY DIAGONALLY UPWARD PULLS (SCHEME DESCRIBED BY SCULLY, PROWLER, AND ANDERSON) (Scheme 19.1)

This scheme is described, in part, on p. 10 of the Book *The Fuel Savers* (Ref. S-60) by Scully, Prowler, and Anderson. The scheme has some defects, listed in a later paragraph.

The scheme employs two curtains: one serves the east half of the window and the other serves the west half. Each curtain is secured to (sealed to) the top of the window frame and also to the pertinent side member of the frame. A canopy, or housing, encloses the top of the curtain system and also the east and west sides. The bottoms of the curtains are weighted so that whenever the curtain is closed the

bottoms press against the floor, forming a fairly good seal.

To open: Pull strings that cause the central regions of the curtains to move diagonally upward, away from the vertical centerline of the window.

### Limitations

Each curtain, being only about 1/8-in. thick, cuts heat-loss only moderately.

The seal along the vertical centerline is likely to be poor, especially near the bottom, unless one curtain has been carefully arranged to overlap the other.

In the open position the curtains appear bulky and awkward.

The lower ends of the curtains, which touch the floor and drag along it, may soon become dirty, and they interfere with sweeping the floor. If they do not rest on the floor, there is no good seal and much convection of air may occur.

The housing, with a total length of the order of 20 ft for a window that is about 6 ft × 6 ft, is bulky and costly.

The requisite area of curtain material is about 25% greater than the area of the window proper. This adds to the cost.

*Scheme 19.1a*

Use two or three cloths in series; or use a quilt. This increases the heat-saving.

*Scheme 19.1b*

Use a 1/4-inch-thick, flexible, solid-foam sheet instead of one of the several layers of cloth. This increases the heat-sving considerably, but also adds to the bulk.

### CURTAIN AND LINER THAT CAN BE SLID HORIZONTALLY (SCHEME EMPLOYING LINER SOLD BY WIND-N-SUN SHIELD, INC.) (Scheme 19.2)

For many decades heavy curtains, and also curtain and liner combinations, have been in common use as window accessories. Usually the liner is between the curtain and the window and may be attached to the curtain or attached directly to the same horizontal rod that supports the curtain. The main function of most liners is to protect the curtain from intense sunlight. Another function, important in summer, is to reflect solar radiation back outdoors and thus help keep the room cool.

Recently, stress has been placed on using liners for the additional purpose of reducing heat-loss via windows on winter nights. Liners for this

purpose are made by Wind-n-Sun Shield, Inc., of Melbourne, Florida. One kind, called *Reversible Piggy-Back* and said to consist of metallized (aluminized) polyester on vinyl is used with the aluminum face toward the outdoors in summr and toward the indoors in winter. The other side is white vinyl. The liner is about 0.005 in. thick, is lightly embossed (textured), and costs about $0.80/ft² at retail.

Users are urged by the manufacturer to staple or tape the pertinent edges of the liner and/or curtains to the sides of the window frame and to employ clips to close the gap at the center.

To open: Grasp the curtain and liner and pull them horizontally away from the vertical centerline of the window. Or pull a pull-cord.

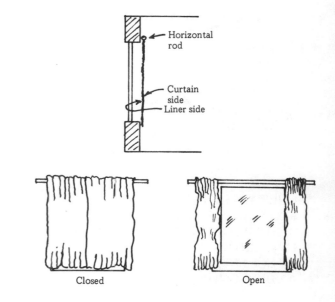

Closed                          Open

*Scheme 19.2a*

As above, except use a hinged insulating lid to cover the gap between the wall and the top of the closed curtain. Often there is a gap of several inches between the wall and the adjacent valance or adjacent top of a heavy curtain. A special lid invented by D. B. Gerdeman of Las Vegas, New Mexico (not Nevada) and covered by his Patent 4,167,205, includes a horizontal rigid strip to which a thick compliant layer of insulating material is attached on the underside. The strip is af-

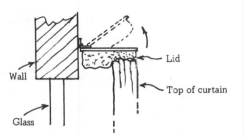

fixed to the wall by hinges and may be raised or lowered by means of a cord and pulley. When down, the lid seals the gap in question. When up, it permits the curtain to be slid freely along its support rail.

## CURTAINS THAT ARE PERMANENTLY AFFIXED AND SEALED TO SIDES OF WINDOW AND CAN BE SLID ALONG A HORIZONTAL BAR AT TOP AND THEN SEALED AT CENTER AND BOTTOM BY MEANS OF MAGNETS (SCHEME DEVELOPED BY CONSERVATION CONCEPTS LTD.) (Scheme 19.3)

This device, somewhat similar to one described in a recent book by E. Eccli (E-60), was developed and produced by Conservation Concepts Ltd. (C. F. Moorhead et al.) of Stratton Mt., Vt. It is called *WARM-IN Sealed Drapery Liner.*

Two curtains (or curtain liners) are used, one along the east side of the window and one along the west side. Each is quiltlike: it consists of a kind of bag of closely woven cloth and, inside the bag, a sheet of bubble plastic (polyethylene with a special coating).

Each curtain is permanently affixed and sealed by tacks to one side of the fixed frame of the window. The upper edge of each curtain carries a row of hooks that engage a horizontal bar and can slide along it. When the two curtains are closed, a vertical edge of one overlaps, and is affixed to, the corresponding edge of the other; seal is provided here by strips of magnetic material 0.06-in. thick. The bottom edges of the curtains are affixed (when the curtains are closed) to the bottom member of the fixed frame of the window. Here too magnets are used. Typical thickness of the region of trapped air between the curtain and window is 2 in.

## Comment

The system provides fairly high R-value, thanks to (1) the use of quilts, (2) the region of trapped air behind the quilts, and (3) the seals.

To open: Pull the two control cords provided. This forces the magnet pairs to release and draws the curtains apart. Alternatively, one may grasp the overlapping edges of the curtains and separate them manually.

Cost: About $2/ft² for liners and sealing devices. Costs of basic curtains, horizontal bar, housing, etc., are not included.

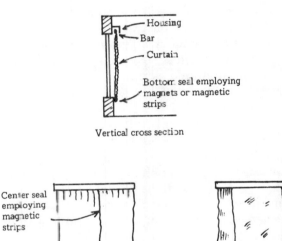

fixed to the wall by hinges and may be raised or

## Limitations

The cost and labor involved in installing the horizontal bar and canopy and applying the hooks and magnets may be high. (If the bar and canopy are already present, the incremental cost of completing the system is much less.)

### Scheme 19.3a

As above, except use Velcro instead of magnets if the window is small. If it is large, the labor involved in joining the Velcro strips each evening and separating them each morning may be excessive.

## CURTAINS THAT ARE PERMANENTLY AFFIXED AND SEALED AT SIDES OF WINDOW FRAME AND CAN BE SLID ALONG A HORIZONTAL BAR AT TOP AND CAN BE SEALED AT BOTTOM BY A LONG HEAVY PIPE (Scheme 19.4)

At the top of the window there is a horizontal bar along which two quilt-like curtains may be slid horizontally. Each curtain is affixed permanently, snugly, to one side of the window frame. Each hangs down 6 in. below the base of the window.

When the curtains have been pulled together (closed) sufficiently so that they overlap about 6 in., the overlap region rests against a vertical aluminum strip that bulges slightly toward the room; accordingly the overlapping regions of the curtains tend to press against each other (and against the aluminum strip), producing a fairly good seal here.

To provide a seal at the bottom, one raises a free-hanging, horizontal 10-lb pipe or bar and installs it so that its ends rest within wedge-shaped hangers that are permanently affixed to the wall (below the window and off to the sides). The pipe presses the lower portions of the curtains firmly against the wall. Thanks to the steep wedge shape of the hangers, the 10-lb pipe exerts a 30-lb sealing force.

To open: Disengage the pipe; raise it, move it north a few inches, and lower it—clear of the hangers. Then open the curtains in normal manner.

### Limitations

The system is generally complex.

One must install a central vertical bar that remains in place throughout the winter.

One must stoop to reach the horizontal pipe.

Appearance may be unattractive.

*Scheme 19.4a*

As above, except hang the heavy horizontal pipe from two very long strings that are attached to the wall at head height. Incorporate hand-grip knobs in the string at a height of 4 ft. above the floor. Then, to lift the pipe, merely grasp the hand-grip knobs and lift them. No need to stoop!

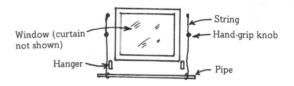

*Scheme 19.4b*

As above, except add, on the wall area adjacent to the heavy horizontal pipe, a compliant, horizontal sealing strip—to improve the seal here and permit using a pipe of only half the weight.

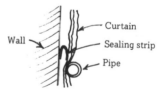

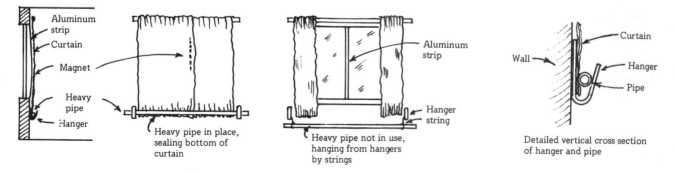

*Scheme 19.4c*

As above, except (per a suggestion by J. C. Gray) use just one big curtain—a single curtain that can extend all the way across the window and overlap the wall on the far side a few inches. Here install a vertical filler strip, to form a rough seal (at this far side). The heavy bar makes the seal at the bottom and also helps make a seal along the filler strip.

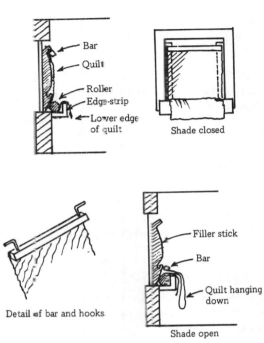

Shade closed

Detail of bar and hooks

Shade open

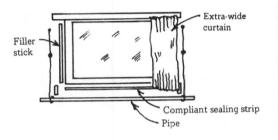

## INDOOR SHADE HELD UP BY HOOKS AT THE TOP (Scheme 19.5)

The upper edge of a quilt (or set of cloths) is permanently attached to a horizontal bar equipped with upward-projecting hooks at the ends. The lower portion of the quilt passes beneath a horizontal roller (4-lb log of wood 2 in. in diameter) confined between the window and an edge-strip attached to the sill. At each side of the window there is a vertical filler strip with a curved face, beveled top, and a notch near the bottom.

To close the shade, one grasps the horizontal bar, raises it high, and presses it against the upper part of the window so that the hooks engage the beveled tops of the filler strip. One then gives a tug to the lower end of the quilt (which projects from beneath the roller) to make it tight in order that the edges will press against the filler strips and thus form edge seals. The weight of the roller insures a bottom seal.

To open the shade, one grasps the horizontal bar, frees it, lowers it, and secures it to the notches in the lower portions of the filler strips. A large loop of quilt then hangs downward below the sill, but does not quite reach the floor.

## Comment

Although simple to construct and install, the system performs well because of the good seals at sides and bottom. The parking problem is solved by lowering the upper part of the shade and allowing it to droop down below the sill. If the window is 4-ft. high, the shade droops about 1½ ft. below the sill.

## Limitations

The cut in heat-loss is modest.

The appearance is somewhat poor.

Short persons may not be able to reach high enough to install the bar at the tops of the filler strips

The log may fall onto the floor.

*Scheme 19.5a*

As above except:

1. Install, on the north side of the shade, an aluminized sheet

2. Attach a stiff handle to the bar; the handle extends downward 1 ft. so that even a short person can reach it.

3. Tie each end of the log loosely to the window frame by means of strings.

## INDOOR MULTI-CLOTH-SHEET THAT CAN BE COLLAPSED DOWNWARD (Scheme 19.6)

This scheme is a modification of a design that was made in about 1977 by C. G. Wing and is said to have performed well.

Several thin, limp cloths are attached at the top to a horizontal bar and are attached at the bottom to a wooden strip that is fastened to the floor just beneath the window. The horizontal bar can be hoisted by means of a cord and pulley, and the various sheets of cloth (and the air spaces between them) then reduce heat-loss through the window. At the start of a sunny day the resident releases the cord and the cloths collapse downward into a long, slender heap on the floor.

### Comment

The shade is easily constructed and performs moderately well.

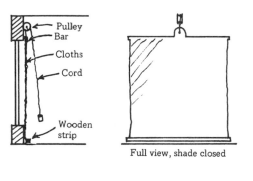

Full view, shade closed

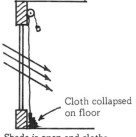

Shade is open and cloths are on floor

## Limitations

The reduction in heat-loss is small because far-IR radiation from objects in the room is absorbed rather than being reflected, and because there are no tight seals at the sides.

The cloths collapsed on the floor look unattractive, take up some floor space, and interfere with sweeping the floor.

*Scheme 19.6a*

As above, but include, on the north side of the shade, an aluminized cloth. This helps by reflecting far-IR.

*Scheme 19.6b*

As above, but install slightly curved filler sticks at each side, so that when the shade is hauled up tight, it will press firmly against these sticks. Thus air leakage at the sides is stopped.

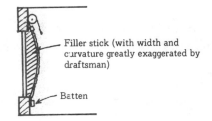

Filler stick (with width and curvature greatly exaggerated by draftsman)

Batten

*Scheme 19.6c*

Instead of using filler strips, use spring-loaded vertical clamping strips such as have been employed by C. G. Wing. Dynamically, the clamping strips are bi-stable: they tend to stay fully open or fully closed (pressing shade edge against wall).

## INDOOR QUILT THAT CAN BE ACCORDION-FOLDED (SUN QUILT) (Scheme 19.7)

This shade, developed by T. K. Price of Sun Quilt Corp., Newport, N.H., is intended for use on very-large-area windows—vertical or sloping—of living rooms or greenhouse areas.

The quilt itself consists of three layers which together are about 1½-in. thick. The south layer is polyester fabric. The intermediate layer is duPont

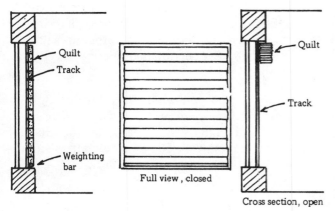

Quilt
Track
Weighting bar

Full view, closed

Quilt
Track

Cross section, open

Fiberfill. The north layer is nylon tricot fabric that has an aluminized face protected by a thin plastic film.

There are horizontal seams 16 in. apart, and along both vertical edges and along the bottom there is a 3-inch-wide band, or roll, of quilting that seals the edges. As the shade is raised or lowered its lateral edges are guided by pairs of rollers (secured to the quilt) that engage vertical tracks that have I-shaped cross section. Within the lower edge of the quilt there is a weighting bar: a 1/2-inch-diameter steel rod.

Operation is fully automatic. A small electric motor is used and is controlled by signals from a photoelectric cell.

In summer the control system is modified so that the shade remains closed during hot sunny periods and is open during other daytime periods.

Cost: About $8/ft$^2$ including tracks, controls, etc., but not installation.

*Scheme 19.7a*

Use the slightly modified design, called Roman Shade II, developed by the Center for Community Technology of Madison, Wisconsin. The shade, which includes 1½ in. of polyester fiberfill (Celanese Corporation's Polarguard), is hoisted manually by pulling on a set of nylon cords. The edges are sealed against the fixed frame of the window by vertical clamping bars served by spring-type hinges. For a detailed description of this shade see *Solar Age*, Aug. 1979, p. 24, or write to Center for Community Technology.

## INDOOR HONEYCOMB-TYPE SHADE THAT IS GUIDED BY VERTICAL SPLINES AND CAN BE HOISTED (SCHEME DEVELOPED BY J. J. ANDERSON, JR.) (Scheme 19.8)

This scheme, developed by J. J. Anderson, Jr., of Ramsey, N.J. about 1977, employs a large honeycomb made of thin plastic (or paper or fabric). The top of the honeycomb is permanently attached by adhesive to a fixed horizontal strip at the top of the window recess. The bottom of the honeycomb is weighted by a horizontal bar which, when the shade is lowered (closed), rests on the sill. Each vertical edge of the assembly is notched so that it can engage, and slide vertically along, a vertical spline that serves as a guide and an edge seal. As the honeycomb is lowered (by releasing the two nylon control cords), its thickness decreases from 3 in. to 2 in.; also the widths of the notches decrease slightly, thus improving the tightness of seal against the splines.

To open: Hoist the shade by pulling on two nylon cords.

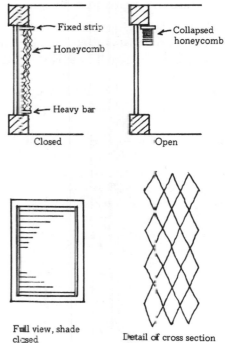

Fixed strip
Honeycomb
Heavy bar

Closed

Collapsed honeycomb

Open

Full view, shade closed

Detail of cross section of honeycomb

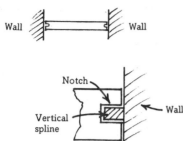

Detail of one end of shade

## Comment

The device is highly effective thermally, thanks to the existence of such a thick region occupied by slender horizontal pockets of trapped air. Yet it collapses to a very small size. As the shade is lowered, the notches in the ends become narrower and hug the fixed splines more closely, resulting in tighter seal.

## INDOOR DEVICE EMPLOYING MANY SLENDER VANES (Scheme 19.9)

Suppose one has a venetian blind that closes tightly, and suppose that each vane is aluminized on both faces. Clearly such a device would provide roughly the same reduction in heat-loss that a sheet of aluminum foil would provide.

The difficulty, obviously, is that venetian blinds ordinarily do not close tightly. When such is the case, the reduction in heat-loss may be small.

It is possible to design a venetian blind that does indeed close tightly. The design may call for springy "edge fingers," or compressible felt, etc. But these features add to the cost and may not be durable.

Fairly tightly closed
venetian blind

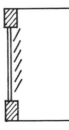

Non-tightly closed
venetian blind

If a great effort is to be made to construct a venetian blind that will be tightly sealed, it may pay to use just a few large vanes, rather than dozens of small ones. (See section dealing with Skylid.)

*Scheme 19.9a*

Use the specially formed, aluminized, tightly-closing, 3½-inch-wide vanes developed by H. P. Shapira and P. R. Barnes of Solar and Special Studies Section, Oak Ridge Associated Universities (Ref. S-145).

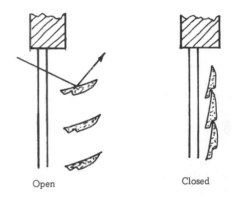

Open                    Closed

## INDOOR DEVICE EMPLOYING BIG, THICK VANES (SKYLID) (Scheme 19.10)

This device, designed and produced by Zomeworks Corp. (S. C. Baer et al.), is well suited to large sloping windows, clerestory windows, skylights, etc. It has the trademarked name *Skylid*.

In a typical application the device consists of a pair of parallel vanes, each running east-west. Each is 4- to 10-ft long and about 20-in. wide. Each has an oval cross section and is 4-in. thick at the center and much thinner near the edges. Each is faced with sheets of reflective aluminum and is filled with fiberglass. The pair of vanes is installed in a large rectangular frame close to the window or skylight in question. Springy fins provide the necessary seals.

Each vane, held by pivots at the ends, may be turned through an angle slightly exceeding 90 degrees. During winter days the vanes are tilted upward to the south to admit most of the solar radiation that has passed through the window. Because the exteriors of the vanes are of shiny

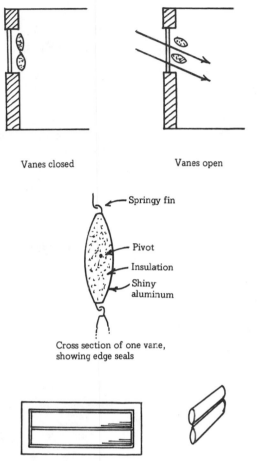

Vanes closed    Vanes open

Springy fin

Pivot

Insulation

Shiny
aluminum

Cross section of one vane,
showing edge seals

Full view of two vanes    Perspective view
and framing    of two vanes

and (because the canisters are attached to the vanes) the torque is transmitted to the vanes—whereupon they turn, i.e., open. At the end of the day a reverse process occurs: there is a back-flow of R-12, the torque reverses, and the vanes close.

Manual override is provided. Thus one can open or close the vanes at any time irrespective of time of day or extent of overcast.

When the vanes are closed a reasonably good seal is achieved. The closing torque provided by the pair of canisters is small, but along each edge of each vane there is a thin, springy sealing fin that engages a fixed channel integral with the fixed frame of the window or skylight.

*Scheme 19.10a*

As above, except replace the thick, tapered vanes with flat vanes that are 1½-in. thick and are made of kraft-paper honeycomb faced with aluminum sheets.

Perspective view of two flat
rotatable vanes

aluminum, little radiation is absorbed; because the vane edges are tapered, little radiation is reflected back to the outdoors. At the end of the day the vanes are closed.

Operation is automatic. For each pair of vanes, one solar-powered actuating system is provided. This includes two steel canisters that contain a material that may be in liquid phase or vapor phase, depending on its temperature and pressure. The material is Freon R-12, a commonly used refrigerant. One canister (painted black) is mounted so as to receive direct solar radiation; the other (silvered) is shaded. When the former is strongly irradiated it warms up and much of its contents is expelled (by the increased gas pressure), via a small copper tube, and transferred to the other canister. Accordingly one canister becomes lighter and the other becomes heavier, with the consequence that gravity exerts a torque on the system

## INDOOR SHUTTER EMPLOYING TWO PLATES, ONE OF WHICH IS HINGED AT THE BOTTOM (Scheme 19.11)

There are two insulating plates, each of 1-inch-thick Thermax. The plate for the upper sash is not permanently attached; it is held in place by clips or the equivalent (see Chapter 13). The plate for the lower sash is permanently attached at the bottom by means of hinges, e.g., hinges consisting merely of strips of tough tape. This plate can be swung open: one merely grasps the upper edge and pulls. It can be opened 45 degrees only because its travel is limited by two cords, one at each side. Each cord is attached to a sheet-metal clip that is taped to an upper corner of the plate. Each cord runs through a screweye attached to the window jamb and extends downward below the sill. the lower end of the cord is attached to a 1/4-lb weight (small block) of wood or very small piece of iron). There is a large knot in the cord at such position that when the plate is opened 45 degrees the knot strikes the screweye, preventing further travel. The weights are sufficiently light that, when the plate is open, they do not close it; but they are heavy enough so that when the plate is nearly closed they close it firmly. Notice that the weights are always below the sill, i.e., never in the way and seldom seen. the cord should be smooth and flexible; nylon fish-line may perform well. Dental floss may be even better.

In opening the shutter in the morning, one removes the upper plate with one hand, opens the lower plate 45 degrees with the other hand, then rests the upper plate on the sloping lower plate.

Note: If the resident finds it difficult to fit the upper plate between the two cords, he may rotate this plate 90 degrees and insert it *endwise* on the lower plate. I assume that the upper plate's width exceeds its height (which can always be arranged by cutting off an upper strip of the upper plate and leaving this strip permanently attached at the top of the upper sash, where it may be largely hidden by the roll-up shade there).

## Comment

This scheme, besides greatly reducing heat-loss, solves the problem of where to store the plates during the day. The system costs little and is easy to install. Note that on a sunny day the parked aluminum-faced plates reflect much solar radia-tion toward the ceiling and thus help heat the room.

## Limitations

When the shutter is open, a large fraction of the lower half of the window is still obscured. However, if one walks up close to the window, one can see over the parked plates fairly well; or one may look around them, obliquely from the side.

*Scheme 19.11a*

As above, except dispense with the hinges. Merely "trap" the lower edge of the lower plate by means

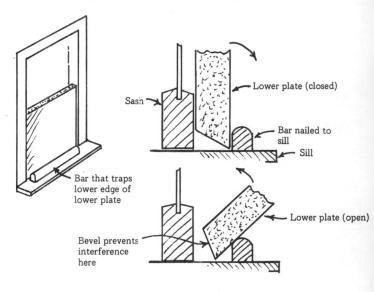

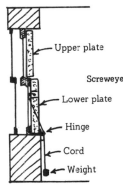

Vertical cross section looking west, with shutter closed

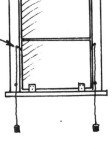

Full view, looking south, with shutter closed

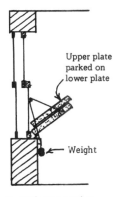

Vertical cross section with shutter open

Location of clips on lower plate

Detail of clip

of a wooden bar nailed to the window sill. Preferably the lower edge of the plate should be tapered to insure that the plate will swing open or closed easily.

Elimination of hinges saves money and labor.

## INDOOR SHUTTER EMPLOYING A PLATE THAT IS HINGED AT THE BOTTOM AND, WHEN SWUNG DOWNWARD, FORMS A WORKTABLE (Scheme 19.12)

This shutter was developed in 1976 by G. R. Hill and Rodney Kipp of Cooperative Extension Association of Clinton County (of New York State). It is described in their pamphlet called "Shut Out the Cold."

The thick insulating plate is faced with plywood and framed with wooden strips. When swung upward (closed) it is held by buttons at the top. When swung downward (opened), it becomes a horizontal worktable. The single leg is of folding type.

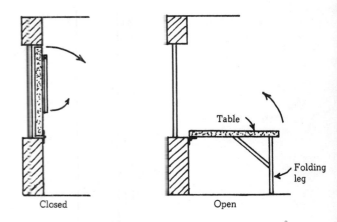

*Scheme 19.12a*

As above except employ two panels, one above the other, joined by hinges. The lower panel swings down and forms a table top, and the upper panel swings down farther to serve as a broad leg.

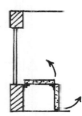

# CHAPTER 20

# DEVICES FOR SPECIAL APPLICATIONS

- Outdoor Hinged Reflecting-and-Insulating Cover for Sloping Window (Scheme 20.1)
- Indoor Hinged Reflecting-and-Insulating Plate for Upper Region of Greenhouse (Scheme 20.2)
- Outdoor Shutter That Serves a Sloping Window and Is Closed by Being Slid Upward Along the Window (Scheme 20.3)
- Outdoor Shutter That Serves a Sloping Window and Is Closed by Being Unfolded and Moved Upward Along Window (Scheme 20.4)
- Between-Glazing-Sheets Flexible Pad That Serves a Sloping Window and Is Closed by Being Unfolded and Slid Upward (Scheme 20.5)
- Indoor Plate That Serves a Sloping Window and Is Closed by Being Slid Downward (Scheme 20.6)
- Indoor Hinged Pair of Plates That Serve a Tall Sloping Window and Are Folded Up and Swung Out of the Way During the Day (Scheme 20.7)
- Indoor Hinged Plate Serving a Rooftop Monitor and Skylight (Scheme 20.8)
- Devices That Serve a Trombe Wall
- Devices That Help Keep the Rooms Cool in Summer

Here I describe some devices that are designed for special applications, i.e., for use with sloping windows, greenhouses, skylights, Trombe walls, etc. Also devices intended to help keep rooms cool in summer are discussed.

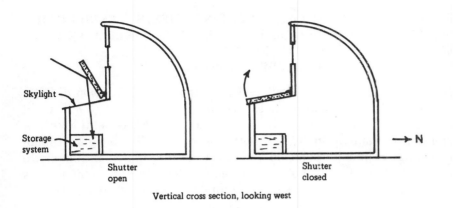

Vertical cross section, looking west

## OUTDOOR HINGED REFLECTING-AND-INSULATING COVER FOR SLOPING WINDOW (Scheme 20.1)

This scheme, or one much like it, was used successfully by Michael Jantzen in a solar-heated house in Carlyle, Ill.

The lower surface of the thick insulating cover consists of aluminized mylar backed by canvas and plywood. The hinges are on the upper edge. When the device is in up position it reflects solar radiation downward through a skylight and toward a storage system. When it is in down position it lies close against the sloping skylight, insulating it. The device is controlled manually from indoors, by means of ropes.

## INDOOR HINGED REFLECTING-AND-INSULATING PLATE FOR UPPER REGION OF GREENHOUSE (Scheme 20.2)

This scheme has been used successfully by R. E. Maes and others of the Environmental Research Institute of Michigan, at Ann Arbor, Michigan. (See *Solar Age* June 1979, p. 20, also Reference I-402f, p. 691.)

The greenhouse is of A-frame type, with a horizontal (largely open) frame at about mid-height. The insulating-and-reflecting plates are of Technifoam, i.e., isocyanurate foam with aluminum foil on both faces. Typical dimensions of plate: 12 ft × 4 ft × 2 in. The edges (not covered with foil) are painted, to prevent deterioration. The south ends of the plates are swung up and down at the start and finish of each day. The north ends, "trapped" by special stop-strips, stay practically motionless. There are no hinges. Nylon ropes 8 ft apart are used to raise the south ends of the plates, which move in unison like a single superplate. The nylon ropes pass over pulleys and are fastened to a single master cable of steel. This is pulled by a hand-operated winch. The operation takes a few minutes. At night the operation is reversed: the south ends of the plates are lowered and the plates become horizontal.

There is a shallow (2-inch-deep) water-filled pond immediately below these horizontal plates. The water is supported by a polyethylene sheet that rests on a coarse wire mesh.

When in up position, the plates reflect solar radiation downward toward the shallow transparent pond and much radiation passes through and reaches the plants below. When in down position, the plates thermally isolate the upper region from the lower region, thus helping keep the lower region warm at night.

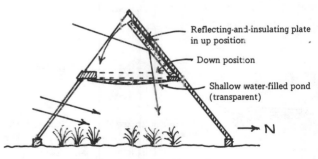

Reflecting-and-insulating plate in up position

Down position

Shallow water-filled pond (transparent)

Vertical cross section, looking west

## OUTDOOR SHUTTER THAT SERVES A SLOPING WINDOW AND IS CLOSED BY BEING SLID UPWARD ALONG THE WINDOW (Scheme 20.3)

This scheme, developed by H. S. Reichmuth and J. L. Barnes of Bear Creek Thunder, Ashland, Oregon, is described in an article entitled "Passive Solar Heating Systems of the Interior Concentrator Type" and published in the *Proceedings of the 1977 Annual Meeting of the American Section of ISES* June 6–10, 1977, at Orlando, Florida. See Vol. 1, p. 11–26.

A single, weatherproof, insulating plate is used. It covers the sloping skylight area, which is near the ridge of the roof and is single glazed. When not needed, it is manually slid down along the roof by a rope and pulley system. At all times it is guided by channels along the east and west sides.

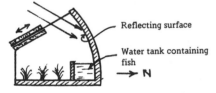

Vertical cross section, looking west

## OUTDOOR SHUTTER THAT SERVES A SLOPING WINDOW AND IS CLOSED BY BEING UNFOLDED AND MOVED UPWARD ALONG WINDOW (Scheme 20.4)

This scheme, developed by H. S. Reichmuth and J. L. Barnes of Bear Creek Thunder, Ashland, Oregon, is described in the same article mentioned in the previous section.

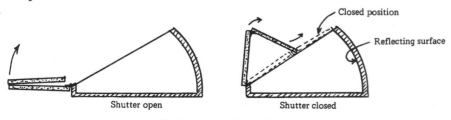

Shutter open        Shutter closed

Vertical cross section, looking west

## BETWEEN-GLAZING-SHEETS FLEXIBLE PAD THAT SERVES A SLOPING WINDOW AND IS CLOSED BY BEING UNFOLDED AND SLID UPWARD (Scheme 20.5)

This device, demonstrated by L. Poisson of Solar Survival, Harrisville, N.H., at the 1977 Toward Tomorrow Fair at Amherst, Mass., employs a 4-inch-thick mattress, or pad, of flexible plastic foam protected by a cover of smooth, free-sliding cloth. The double-glazed window, sloping at 60 degrees from the horizontal, consists of two sheets of Kalwall Sun-Lite about 4 in. apart.

To close the shutter, one pulls an overhead rope which, via an overhead pulley, hauls the pad upward into the space between glazing sheets. To open the shutter, one pulls a rope that is near the floor and runs under a pulley in a storage box. As the pad is drawn into the box, it is constrained to follow a bent, or folded, path and accordingly the box does not have to be very long.

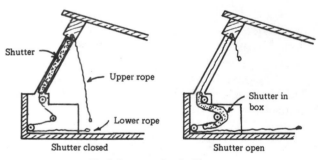

Shutter closed        Shutter open

Vertical cross section, looking west

4-in.-thick pad, foam-filled

## Comment

The shutter is very well protected and it cuts heat-loss greatly.

## Limitations

The box takes up much space. But it can be disguised and used as a window seat.

The Sun-Lite is not transparent. It blocks view of the outdoors. It could be replaced by glass, but this would greatly increase the cost.

*Scheme 20.5a*

Recess the box into the floor; make it an integral part of the floor.

*Scheme 20.5b*

Put the box at the top; integrate it into the ceiling.

## INDOOR PLATE THAT SERVES A SLOPING WINDOW AND IS CLOSED BY BEING SLID DOWNWARD (Scheme 20.6)

I learned of this scheme from David Wright in October 1977.

The heart of the shutter is a rigid insulating plate, supported by tracks, mounted just below a gently sloping skylight. On a sunny day in winter the shutter is hauled upward, parallel to the roof, into a special recess or slot within the roof structure. In the evening the plate is hauled downward and lies close beneath the skylight.

On hot summer days the shutter is hauled upward only 10% of the way, so that gaps occur at the lower and upper ends of the plate. The gap at the top exists because the upper edge of the plate is extra thick and makes a seal only when the plate is fully closed; see sketch.

### Comment

The shutter greatly cuts heat-loss on winter nights, and it is entirely out of the way.

### Limitations

The system must be planned carefully before the roof is built.

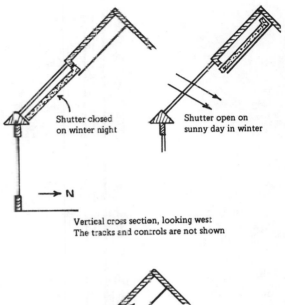

Shutter closed on winter night

Shutter open on sunny day in winter

N

Vertical cross section, looking west
The tracks and controls are not shown

On sunny day in summer the shutter is open just far enough to allow venting of hot air

Accessibility for repair is poor.

Care is needed to insure that the closed shutter is well sealed.

## INDOOR HINGED PAIR OF PLATES THAT SERVE A TALL SLOPING WINDOW AND ARE FOLDED UP AND SWUNG OUT OF THE WAY DURING THE DAY (Scheme 20.7)

There are two rigid insulating plates: a larger one and a smaller one. The upper edge of the larger plate is attached by hinges to the top of the window frame. The smaller plate is attached to the lower edge of the larger one by hinges. A 3-foot-long pull string dangles from the lower south edge of the larger plate.

On winter nights, buttons hold the plates close against the window. In the morning the resident swings the lower plate until it lies close to the north (under) face of the larger plate. Then he

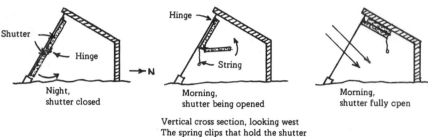

Night,
shutter closed

Morning,
shutter being opened

Morning,
shutter fully open

Vertical cross section, looking west
The spring clips that hold the shutter
open are not shown.

grasps the lower edge of the pair and pulls it north;
then (with the aid of a long stick) he pushes it
upward, high above his head, until it is held by a
spring clip.

In the evening he releases the pair of plates by
pulling down on the dangling strings. He unfolds
them, pushes them against the window, and locks
them tight against the window by means of buttons.

## Comment

Note that the shutter—despite its huge size—is
entirely out of the way during the day. At night it
cuts heat-loss greatly.

## Limitations

If the folded shutter, secured near the ceiling,
were to fall, it might hit someone on the head.
But the (Thermax?) plates are so light that little
harm would be done.

When the lower plate is swung north and up, it
may strike objects that are only a foot or two
above ground level.

## INDOOR HINGED PLATE SERVING A ROOFTOP MONITOR AND SKYLIGHT (Scheme 20.8)

The shutter proper is a 3-in. plate of urethane foam,
both faces of which are covered by Thermoply. The
hinges are at the north edge. The south edge
sweeps upward to allow solar radiation that has
passed through the skylight to proceed onward
and downward into the building.

The device is opened by operation of a small,
boat-type winch which, via a steel cable, turns a

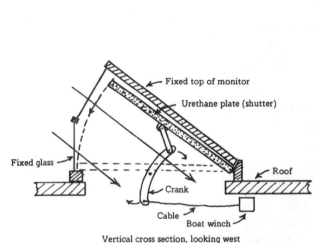

Vertical cross section, looking west
Shutter is raised (open)

crank, or lever, that raises or lowers the insulating
plate.

For further details see the original article by G.
Gerhard (G-140).

*Scheme 20.8a*

As above, except that the insulating plate itself
serves as the top member (roof) of the monitor.

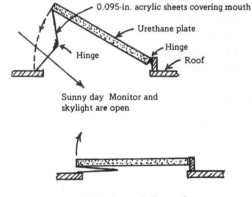

Sunny day  Monitor and
skylight are open

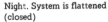

Night. System is flattened
(closed)

That is, the shutter. instead of being within the monitor and below the monitor roof, is the monitor roof. When the shutter is opened, the "mouth" at the south is covered by a hinged pair of 0.095-in. acrylic sheets. When the shutter is closed, the pair of sheets folds up (flattens). At each joint of the glazing system there is a neoprene seal. For further details see G-140.

## DEVICES THAT SERVE A TROMBE WALL

There are many approaches to providing nighttime insulation for a Trombe wall. See, for example,

> Chapter 6 through 10 for various kinds of outdoor shutters.
>
> Chapter 11 regarding between-glazing-sheets devices, including:
>> roll-up aluminized mylar shade with roller at top
>>
>> roll-up aluminized mylar shade with roller at bottom
>>
>> set of cloths that may be collapsed upward
>>
>> set of cloths that may be collapsed downward
>>
>> beadwall system
>
> Chapter 14 regarding indoor devices that slide, including:
>> sliding quilt
>>
>> sliding set of Styrofoam boards

To provide insulation for a Trombe wall that slopes 45 degrees, one may use a scheme developed by D. Munday and described on p. 762 of Reference I-402f. At night there is a 3-inch-thick plate of Styrofoam between the Trombe wall and

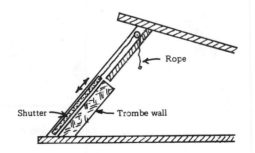

the window. The plate is strengthened by a frame of pine wood and faces of 1/4-in. plywood. At night the plate is hauled up (by means of a 1/2-in. nylon rope running over a pulley) into a recess, or slot, in the upper region of the sloping wall.

## DEVICES THAT HELP KEEP ROOMS COOL IN SUMMER

Because this subject is not part of the main scope of this book, I deal with it only briefly.

Nearly any opaque shutter or shade can be used to help keep south-facing windows cool on sunny summer days: merely keep the devices in place on the windows throughout the day. It is helpful, of course, if the south faces of the shutters or shades are highly reflective, i.e., covered with shiny aluminum foil. A white covering is nearly as effective.

The effectiveness is especially great if the device is mounted outside the window. Then the solar radiation does not pass through the glazing at all. Glazing has some absorptivity, and if solar radiation passes through it (or passes through it twice—coming and going), it becomes hot and contributes to making the room too warm.

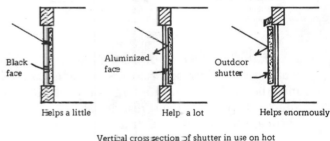

Vertical cross section of shutter in use on hot sunny day in summer

Outdoor venetian blinds, or other special sets of outdoor louvers, can be very effective in excluding solar radiation. Such equipment, called Kool-Shade, is made by Solar Science Industries.

Several persons who use black indoor shutters make excellent use of them in summer in this way: (1) they allow a 2-in. air space between the glazing and the shutter and (2) they provide, at the bottom and the top of the window region, channels to the outdoors. Thus hot air in the 2-in. space travels upward and thence to outdoors and cooler outdoor air enters near the bottom. (In some instances the device is modified for winter use: the above-mentioned channels at the bottom and top of the window are closed and another set of channels extending to the room interior is opened; thus the upward traveling stream of warmed air is discharged into the room, near the ceiling, and cooler air from near the floor enters the 2-in. space.) Reference: Sunset Magazine, April 1977, p. 134, which describes a scheme developed by David Wright. See also *Alternative Technologies*, Issue 3, April-May 1977, for a description of a scheme used by Alan Garinger. See also some schemes employed successfully by R. S. Levine and described in Chapter 15.)

A window so equipped may be used as an integral part of an air-type solar heating system. The airflow may be directed (with the help of a small blower) to a bin-of-stones in the basement.

If an aluminized shutter can be rotated by at least 45 degrees, it can be used in summer to exclude radiation and in winter to direct the incoming radiation toward a small, direct-gain storage system near the ceiling or near the floor. Especially good collection efficiency may be achieved if the shutter is somewhat flexible and can be bent to a roughly cylindrical-parabolic shape. Many such

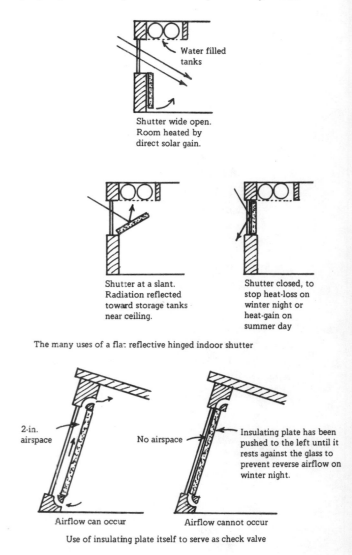

Shutter wide open. Room heated by direct solar gain.

Shutter at a slant. Radiation reflected toward storage tanks near ceiling.

Shutter closed, to stop heat-loss on winter night or heat-gain on summer day

The many uses of a flat reflective hinged indoor shutter

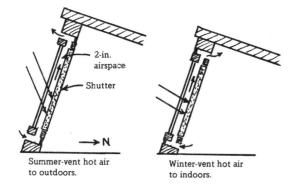

2-in. airspace

Shutter

Summer-vent hot air to outdoors.

Winter-vent hot air to indoors.

If, during the day, or during the night, an unwanted back-flow of air occurs, suitable dampers may be installed to stop this—e.g., a 0.001-in. sheet of polyethylene resting against a near-vertical screen. Another strategy is to temporarily displace the shutter: push it up against the window, automatically stopping all airflow.

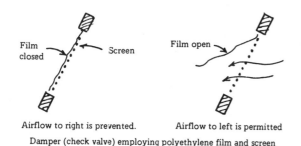

Film closed

Screen

Film open

Airflow to right is prevented.

Airflow to left is permitted

Damper (check valve) employing polyethylene film and screen

2-in. airspace

No airspace

Insulating plate has been pushed to the left until it rests against the glass to prevent reverse airflow on winter night.

Airflow can occur

Airflow cannot occur

Use of insulating plate itself to serve as check valve

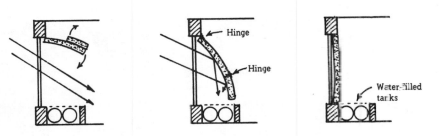

The many uses of a flexible (flat or curved) hinged indoor shutter

schemes are described in my book of Inventions (S-235cc). Two examples are described in the accompanying sketches.

**Partially reflecting glass**  Obviously the tendency for south-facing rooms to become too hot on sunny summer days is reduced if the windows are glazed with glass that reflects (say) 30% or 60% of the incident solar radiation. Several glass manufacturers routinely produce such partially reflecting, partially transmitting glass, and use of such glass in office buildings is becoming common. However, such glass reduces the natural illumination of the rooms and reduces the extent of solar heating achieved in winter.

**Adhesive-coated sheets of partially reflecting plastic**  An alternative approach is to apply a partially reflecting plastic film to ordinary glass glazing. Such films are described in Appendix 7.

# CHAPTER 21

# CONCLUSIONS

- Savings
- Priorities
- Choice of Design Approach
- Buy or Make?
- Woes Concerning Installing Certain Commercial Devices
- Choice of Approaches to Edge Seals
- Prediction Concerning Goals in Seal Design
- Great Appeal of Very Lightweight Insulating Plate
- Fire Hazard Enigma
- The Future of Roll-Up Shades
- Suggestions to Homeowners

## SAVINGS

Shutters and shades could save this country 1 or 2% of its annual purchased energy, and much more if, as many economists claim, the true overall incremental cost of oil is two or three times the published cost.

Most homeowners could save much money and increase their comfort by installing shutters and shades. Applied to single-glazed windows in cold regions, such devices may pay for themselves in 1 to 4 years and may provide a profit of $1/ft$^2$ each year thereafter. Applied to double-glazed windows the payback period is about twice as great, i.e., still acceptably short. In warm regions, inexpensive low-R devices applied to single-glazed windows may pay for themselves in 2 to 6 years, but if the windows are double-glazed the devices must be of extremely-low-cost type if they are to be cost-effective.

The potential savings on apartment buildings, commercial buildings, schools, etc., in cold or moderate climates are very large.

Especially on passively solar heated buildings or other buildings in which the window area is large, shutters and shades may be very cost-effective.

## PRIORITIES

Installing thermal shutters and shades ranks about sixth among the steps one should take to reduce heat-loss from a building:

First: Turn down the thermostat and dress more warmly.

Second: Insulate the attic or roof.

Third: Stop air-leakage at windows, outer doors, attic, and basement.

Fourth: Insulate the walls.

Fifth: Install storm windows.

Sixth: Install thermal shutters or shades.

## CHOICE OF DESIGN APPROACH

People who are considering using thermal shutters or shades face difficult choices, including the choice between:

indoor and outdoor devices

devices that hang, slide, fold, or roll up

devices that consist of thick, strongly built assemblies, or simple lightweight plates, or thin flexible sheets

devices that rely mainly on trapped air, or on thick foam, or on reflective aluminum surfaces

devices that employ formal seals, informal not-so-tight seals, or no seals at all

devices that remain in view and ready for use throughout the year, or throughout the winter only, or throughout winter nights only

devices that are integral parts of the walls and are installed while the house is being built, or are designed for retrofit

devices that supplement existing curtains and shades, or devices that replace them

devices designed mainly for beauty, or mainly for utility.

Whatever the choice of approach, there is no one ideal design of shutter or shade. There are scores of sets of conditions and scores of best designs.

## BUY OR MAKE?

Commercially available shutters and shades have several drawbacks, but also great merit. The cost is usually about $4 to $10 per square foot, and installation by a professional may add 40 to 80% to the cost. Most such devices are conspicuous and some are unattractive. Many are flammable. Some may wear out in 5 or 10 years. Opening and closing the devices each day may be a nuisance.

However, the heat-saving is large, especially as the saving of money will increase each year as costs of back-up energy increase. New kinds of commercially produced devices are appearing each month. Inventors are trying out dozens of new types. There is sure to be much progress in the next few years.

Installing thermal shutters and shades—or making and installing such devices—is a rewarding do-it-yourself activity. A homeowner who is familiar with tools may be able to build and install highly effective devices that cost little.

## WOES CONCERNING INSTALLING CERTAIN COMMERCIAL DEVICES

I have personally installed three high-performance commercial shutters and shades. In each case, the supplier's literature implied that installation was simple. "Merely mark a line where . . . ," "Merely score the material with a sharp knife and break on the line . . . ," "Simply fold, staple, and attach . . . ," "Drill three holes that line up accurately with . . . ,"

Alas, reality was grim. Small errors in layout and markings, small deviations of the knife, uncertainty as to which way to fold the material, the need for three or four hands while stapling, worries as to how to be sure the parts would line up accurately—such matters greatly slowed the work. Other troubles included slight warpage of components, uncertainty as to how to apply the staples and glue, a tendency of some parts to be very slightly oversized and thus inoperative (or undersize with resulting wide gaps), and the need to reread the cryptic instruction sheet over and over.

Using a large battery of tools and many years' experience in building bookcases, kayaks, toys, etc., I succeeded in completing each installation. But the time expended was 3 to 5 hours per window. (I could probably complete a second installation in half the time.)

I conclude that ease-of-installation should be given more attention by shutter designers, instruction sheets should be written more clearly and illustrated better, and claims that installation takes only a half hour or so should be toned down.

Alternatively, the supplier should recommend that a skilled professional do the installing and should warn that this may add considerably to the cost.

## CHOICE OF APPROACHES TO EDGE SEALS

Chapters 17 and 18 have described many ways of sealing the edges of roll-up shades. Some of the main classes of seals may be summarized as follows:

| Location of shade | Sealing equipment used | Sketch (horizontal cross section) |
|---|---|---|
| In recess | None | |
| In recess | Filler strips | |
| In recess | Filler strips and vertical bars | |
| In recess | Simple channels | |
| In recess | Channels that accommodate beaded edges of shade | |
| Overlapping the fixed frame of window | None | |
| Overlapping the fixed frame of window | Filler strips | |
| Overlapping the fixed frame of window | Vertical bars | |

## Some comments:

If the shade is thin and transmits far-IR readily and the window is reasonably tight, it is a waste of time and money to provide good edge seals. Even with them, the shade will cut heat-loss only slightly.

If the shade is thick and stops heat-flow by conduction and radiation, use of high-quality seals may be highly rewarding. Conversely, if the edge seals are very poor, the effectiveness of the thick shade is largely defeated.

If the vertical edges of the shade are beaded (thickened) and travel up and down in tight-fitting vertical channels, the edge seals are fairly tight. But there are these drawbacks:

It is necessary to match exactly the distance between channels to the width of the shade. If the shade is 1/2 in. too wide, it may belly out awkwardly. If it is 1/2 in. too narrow, it may not fit into the channels or may slide with too much friction. If the shade changes in width, e.g., shrinks on being washed, trouble may result.

The channels are cemented or nailed in place and cannot be removed easily when summer comes.

The cost of the channel may be $1 per linear foot.

If the vertical edges of the shade are simple (non-beaded), a fairly tight seal may be arranged by providing vertical channels that are very slender. If the channels are 1-in. deep, the tolerance on the width of shade may be great, 1/2 in. say. But there are these drawbacks:

If the seal is fairly tight, there may be appreciable friction. Raising or lowering the shade may be difficult.

The edges of the shade may become worn or frayed after a few years.

The channels may gradually widen.

The channels are so deep (1 in.) that they block some light. Also they are highly visible and may interfere with removing the window sashes for repair.

The channels are cemented or nailed in place and cannot be removed easily when summer comes.

The cost of the channel may be $1 per linear foot.

Schemes in which the shade is extra-wide, overlaps the walls, and is pressed against them by vertical clamping bars have these advantages:

A fairly tight seal is achieved.

The tolerance on shade width is great. Replacement shades of slightly different width will perform just as well as the original shade.

The window proper is not obscured or encumbered in any way.

The system continues to work well even if the present shade is replaced by a much thicker one or is replaced by two shades in series.

In summer the vertical bars can be removed and stored in the basement.

Some drawbacks are:

An extra-wide shade is required.

Special support brackets for the shade are required.

The system disfigures the wall areas adjacent to the window.

Schemes involving recessed shades, filler sticks, and vertical clamping bars have most of the advantages listed above. The main drawback—disfiguring the window—is slight inasmuch as the special equipment is thin, is mounted within the recess and close to the jambs, and the sticks can be painted to match the jambs.

## PREDICTIONS CONCERNING GOALS IN SEAL DESIGN

I expect that, in the next few years, companies selling thermal shutters and shades will continue to stress the tightness of the seals provided, even if the seals are not very tight. Most people know that heat-loss by leakage of air through windows is a serious matter and many people will be attracted by the claims that certain devices are ". . . tightly sealed, and can stop heat-loss by leakage of air."

But I expect that the long-term trend will be very different. People will learn that the most effective way to stop heat-loss by leakage is to stop it at the source; stop the leaks in the window system itself. To make a shutter so well sealed that it will stop such leakage—yet can be opened and closed easily each day—is difficult and costly. People will learn to focus their attention on achieving shutter or shade seals that are just good enough to stop indoor convective currents, or just good enough to stop such currents and stop condensation of moisture on the windows; such seals are far easier to achieve.

Will the use of vertical channels at the sides of the window grow? Or will simpler schemes be found? I am betting on the simpler schemes.

## GREAT APPEAL OF VERY LIGHTWEIGHT INSULATING PLATE

If a shutter employs an insulating plate that weighs more than about 8 lb, the design of the shutter as a whole perforce becomes highly complicated. Some people may find it difficult to lift such a weight every day; therefore the need to lift it must be avoided. In short, the shutter must be hinged or must slide horizontally. If it is hinged, children might play with it and hang from it, for example; thus the hinges must be strong and the insulating plate must be provided with a strong edging or frame. If the plate slides, it will preempt a large area of wall, unless a special recess was provided when the wall was originally built. Windows may differ slightly in size, but to adjust the size of a fully framed insulating plate is very difficult. In summary, the decision to use a plate too heavy for a frail person to lift each day leads to a train of complications—and considerable expense. Furthermore, once such a shutter has been installed, the homeowner will be reluctant to disturb it when summer comes: he is likely to leave it in place throughout the summer, even if the device is bulky and unattractive and he does not intend to use it again until autumn.

Consider, on the other hand, a shutter employing a 1/2-in. sheet of Thermax. A 36 in. × 36 in. plate of such material weighs only 1 lb. Lifting such a plate is a cinch for anyone. And, incidentally, if such a plate were to fall three feet onto someone's head, it would merely make the person smile. The significance of this is that the plate may merely be secured against the window: clipped into place in the evening and removed and stored at night. No hinges are needed. No strength is needed—no perimeter frame—because no strong force will ever be exerted on the plate. Because the plate has no frame, its size is easily adjusted; for example, a 1/2-in. slice can be trimmed off with a knife or carpenter's saw. Fitting a dozen plates to a dozen windows, each of which is slightly off-square and off-size, is simplicity itself. Throughout the summer the plates may be stored in a closet or basement. The edges of the plates may be taped in any way the homeowner sees fit, and the faces may be painted or otherwise beautified. The entire situation (buying, cutting, fitting, beautifying, removal in the spring) is entirely within his understanding and control.

## FIRE HAZARD ENIGMA

Plates of Thermx, Styrofoam, etc., are flammable. Some of them, on burning, emit highly poisonous gases. The question is: Is the homeowner well-advised to make indoor thermal shutters of such material?

Alas, there is no fully adequate answer to this question. The manufacturers of the products issue statements on flammability and toxicity, but their statements are so technical and so carefully phrased that they provide no clear answer. A very cautious safety engineer might say: "Do not use such materials in the home. They can be very dangerous." Other safety engineers can point out that most houses contain many objects that are flammable, and all such objects, when burning, emit poisonous gases. They may even say that shutters are less likely to catch fire than sofas, rugs, and curtains are.

One may point out also that many such foam

materials contain fire retardants, and rigid foam plates that are faced with aluminum foil are to some extent (or for a short period of time) shielded from a fire by the foil.

For sure, the world will owe a great debt to the first company that develops an insulating plate that has all the desirable characteristics of organic foam plates but without the Achilles heel of flammability.

## THE FUTURE OF ROLL-UP SHADES

I expect that new kinds of roll-up shades will have a big future. Roll-up shades are logical favorites because they are inexpensive, are easy to install, occupy little space, and are easy to operate. If one or two aluminized surfaces are incorporated in such a shade, and especially if two or three sheets are mounted on the same roller and, when un-rolled, define several regions of trapped air, excellent thermal performance will result. The R-value may be 3 to 8 and the heat-saving may be 40 to 70% for double-glazed windows and 60 to 80% for single-glazed windows. The higher values may be realized only if fairly tight edge-seals are provided; hence the large number of edge-sealing schemes described in Chapters 17 and 18.

It seems likely that when a designer decides to incorporate a moisture barrier in a shutter or shade, he will consider employing an aluminum foil or an aluminum coating for this purpose. Such foil or coating will perform two functions: preventing passage of moisture and reflecting far-IR radiation.

Designers of shades employing aluminum surfaces may agree on the merits of having the aluminum flanked on at least one side with air only. They will recognize that the far-IR-reflecting role of aluminum is largely negated if the aluminum is closely flanked on both sides by a fabric, quilt, or foam plate that has high far-IR absorptance and emittance.

## SUGGESTIONS TO HOMEOWNERS

On some very windy day, inspect each window and find whether it is allowing much cold air to leak in. If so, stop the leaks: caulk or weather-strip the windows, or hire an expert to do so.

Read about shutters and shades. Ask advice from friends and from pertinent stores and agencies. Examine the shutters and shades in friends' houses.

Choose a type of shutter or shade. Install just one and see how well you like its appearance and operation.

If you have doubts as to how much it cuts heat-loss, obtain some thermometers and actually measure the reduction. See Appendix 1.

After a few weeks, install more shutters and shades. Consider using a variety of types, according to the kinds of windows, their locations, and their importance for view, ventilation, etc.

Give early consideration to use of roll-up shades that include an aluminized sheet.

Also give early consideration to simple not-permanently-attached insulating plates of Thermax. They are very effective and very easy to prepare and install.

Give attention to devices that can be left in place continually, night and day, i.e., added sheets of glass, transparent plastic, or translucent plastic.

If you propose to buy an elaborate, commercially available device, try to estimate realistically the labor involved in installing it.

Give thought to potential fire hazards, hazards from mechanical injury, etc.

Train yourself and family members to close the shutters and shades at the end of each day in winter.

Inspect the windows and shutters and shades occasionally. See whether the windows are still tight and whether the seals of the shutters and shades are still at least reasonably tight. Remember that a typical shutter or shade can perform far worse—or far better—than its nominal R-value would suggest, depending on the amount of air leakage and the extent to which the shutter or shade reduces it.

# APPENDIXES

# HOW TO MEASURE THE HEAT-LOSS

There are various methods of measuring the heat-loss through a window or through the combination of window and shutter or shade: relative methods and absolute methods. Some methods take air-leakage and wind speed into account, and some do not. I explain first a simple method of measuring the relative amount of heat saved by employing a given shutter on a given double-glazed window; I call it the *between-glazings temperature head method*.

## BETWEEN-GLAZINGS TEMPERATURE HEAD METHOD

Consider a south window that includes regular sashes and storm sashes, with a 3½ in. air space between. Obtain three ordinary thermometers that have been found to read alike in a warm room and also read alike when put inside a refrigerator at about 30°F. (If they do not read alike, make suitable corrections each time readings are made.) Call the thermometers A, B, and C, and mount them as follows:

A. outdoors, 3 in. from the center of the glass of the storm sash

B. between the sashes, and 1/2 in. from the center of the storm sash

C. indoors, 4 ft from the center of the storm sash

At night, with no shutter in place, and with the outdoor temperature such that A reads 30°F and the indoor temperature such that C reads 70°F, read B. Find the temperature head H by subtracting 30 from B's reading:

Temperature head H = (B's reading) − 30°F.

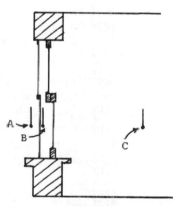

Locations of Thermometers A, B, and C used in determining temperature head H

**The interpretation:** The amount of heat lost through the window if the wind has some standard speed and direction and little air leaks through the window—is roughly proportional to H. Thus by measuring H with and without a given shutter in place, one can find the percent heat saving that the shutter produces.

201

Specifically, the percent heat saving is:

$$\frac{H_{no\ shutter} - H_{shutter}}{H_{no\ shutter}}$$

**Example:** Suppose that, with no shutter in place, thermometers A, B, and C read 30°F, 45°F, and 70°F. Then

$$H_{no\ shutter} = 45 - 30 = 15°F.$$

Suppose that with a certain shutter in place, and installed in a certain way, the readings are 30°F, 33.5°F and 70°F. Then

$$H_{shutter} = 3.5°F.$$

Assuming that the wind conditions were the same throughout and the window leaked little or no air, then:

$$Percent\ heat\ saving = \frac{H_{no\ shutter} - H_{shutter}}{H_{no\ shutter}}$$

$$Percent\text{-}heat\text{-}saving + \frac{15 - 3.5}{15} = 75\%.$$

This is a very effective shutter. A 1/2-in. plate of Thermax installed close to the sash (and with edge gaps not exceeding 1/8 in.) may provide such performance.

Usually the outdoor temperature will not be exactly 30°F and the indoor temperature will not be exactly 70°F. The procedure then is to normalize the actual temperature readings, i.e., adjust all three of them in the simplest linear manner so as to make the A and C readings exactly 30°F and 70°F.

**Example**  Suppose the actual readings are 15, 19, and 65.

Subtract 15 from each, to obtain         0, 4,    50.
Multiply each by 40/50, to obtain:       0, 3.2,  40.
Add 30, to obtain:                       30, 33.2, 70.

The extreme values are 30 and 70, and the head H is 3.2.

**Extreme case**  If the indoor shutter were ideally effective—stopped all heat-flow—then thermometers A and B would read alike. The head H would be zero, and the saving would be 100%. This con-

clusion is furnished by common sense and also by the formula stated above.

## OUTDOOR GLAND METHOD

Here Thermometer B is mounted in a tiny half-box, or gland, that is taped in airtight manner to the outer face of the glazing. The box is made of 1/2-in. Thermax or the equivalent. The lower the reading of this thermometer, the greater the heat-saving the indoor shutter or shade provides. Here too it is helpful to normalize the data to indoor and outdoor temperatures of 70°F and 30°F.

The reading of Thermometer B has no absolute significance: the experimenter cannot compute, from this reading, the percent heat-saving provided by the shutter under test. But often he is content merely to compare several shutters. If one of these has a known absolute value of percent saving, then the absolute values of the others can be estimated.

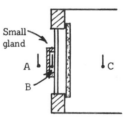

For a window such as Thermopane, where the between-glazings space is small and inaccessible, this method of evaluating the thermal performance of a shutter or shade may be the best choice.

## INDOOR GLAND METHOD

Here Thermometer B is mounted in a gland that is taped to the north side of the north glazing. The gland is between the glazing and the shutter or shade. The lower the reading, the greater the heat saving provided by the indoor shutter or shade.

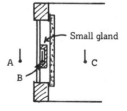

## METHOD APPLICABLE TO OUTDOOR DEVICES AND BETWEEN GLAZING-SHEETS DEVICES

The indoor gland method may be used when the insulating device is situated outdoors or between the glazing sheets. Here, obviously, the heat saving is greater the higher the reading of Thermometer B.

## METHOD APPLICABLE TO SINGLE-GLAZED WINDOW

The above-described outdoor gland method and indoor gland method may be used when the window is single glazed.

## CRUDE ABSOLUTE METHOD FOR USE IN LABORATORY

### The Equipment

One employs a six-sided box, one face of which (Face 1) is slightly larger than the window to be tested.

Face 1 is extremely well insulated, e.g., with 12 in. of urethane foam.

The other faces are insulated at least fairly well, e.g., with 6 in. of urethane foam.

A long-stem (12-inches-long) thermometer is employed. The bulb is inside the box and the graduated stem is outside.

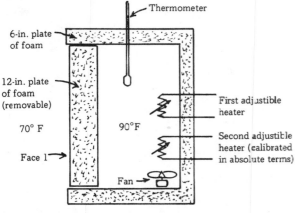

Cross section of test box used in measuring the absolute heat-loss through a window. (The window is to be installed in place of the 12-in. plate of foam.)

A tiny electric fan within the box keeps the air in motion so that the temperature there is everywhere the same.

A tiny, variable-output electric heater is included and is adjusted so that, with the ambient temperature 70°F, the temperature inside the box holds steady at 90°F. The tiny heater provides just enough heat to replace the heat that leaks through the six faces of the box.

A second tiny, variable-output electric heater is included—to be used as explained below. This device is calibrated in terms of Btu/hr.

### Procedure for Window Alone

Remove Face 1 and install, instead, the test window. Be sure it is sealed so well that no air can leak through it or around it.

Adjust the second heater until the temperature again holds constant at 90°F, and record the power output of this device. This is the amount of heat leaking through the window. (To be strictly accurate a small correction should be made to take into account the amount of heat that was lost through the 12-inch plate of urethane foam used earlier. But this quantity is very close to zero. Ordinarily, no correction is needed.)

### Procedure for Window plus Shutter

Install the given shutter on the window.

Again adjust the second heater until the temperature holds constant at 90°F. Record the device's power output. This is the heat-loss through the window-plus-shutter, with no leakage of air through the window.

### Heat-Saving Attributable to Shutter

Comparing the results without and with the shutter, one can find the heat saving attributable to the shutter. This can be expressed in absolute terms (Btu/hr, or Btu/hr per square foot) or in relative terms (percent heat saving). Warning: the result applies only when (1) the shutter is applied just to this type of window, (2) the shutter is installed just in the present way, and (3) the window leaks no air

**Example**   With a certain window in place (serving as Face 1), the power output of the second heater is 120 Btu/hr.

When a certain shutter is added, the power output is half as great: 60 Btu/hr.

One concludes that the shutter (installed in this particular way in this particular window—which leaks no air) cuts the heat-loss by 60 Btu/hr. In relative terms it provides a percentage heat saving of 50%.

## OTHER METHODS

Various prestigious laboratories employ more elegant equipment.

Usually there are two boxes, I understand: one hot, the other cold, the window being mounted between them.

Usually there is a "guard band" along the perimeter of the window, i.e., an edge-interface system that either eliminates, or compensates for, unwanted edge effects.

Usually the temperature sensors are more elaborate. Recorders of various types are used.

In some instances there is provision for simulating wind on one face of the window.

## Discussion

I have little enthusiasm for these equipments, because:

1.  They are expensive.
2.  Although they gain in accuracy, they lose by having less resemblance to real-life situations (actual windows in actual houses located near actual windbreaks and exposed to actual winds). They are at their best in unrealistic situations: windows that leak no air, shutters perfectly sealed, uniform simulated wind.
3.  The results obtained for certain simple, well-sealed shutters can be found more simply by calculation, using the known R-values of the insulating plates used.

# SEALS:
# SOME EXPERIMENTAL RESULTS

Having found no adequate accounts of how tight the seals at the edges of shutters and shades should be, or how large the penalties are if the seals are not tight, I made some crude experiments early in 1979. The results are summarized below.

## FACE SEAL

To find the tolerance on face-seal gap, I conducted several experiments on an east window of my house in Cambridge, Mass. The window is double-glazed: it has a main window and, 3½ in. from it, a storm window. I obtained a 1½-inch-thick sheet of Styrofoam SM and mounted it at various distances (0 in., 1/32 in., 3/16 in., and 5/16 in.) from the main window glazing. Small shims were used to define the spacings. In each case there was an edge gap of 1/8 in. between the edge of the plate and the adjacent member of the sash frame. For each position of the plate, I measured the actual percentage of heat-saving achieved by the plate. The experimental method of evaluating the saving is indicated in Appendix 1.

The results are shown in the accompanying graph. To my surprise, and contrary to rumor, the heat-saving remains very high even when the face-seal gap was increased to 5/16 in. The same general result was found using a 1/2-in. plate of Thermax.

· I conclude that even when the thickness of the air space between the insulating plate and the glazing is as great as 5/16 in. (and even when there is a 1/8-in. edge gap at each edge of the plate), practically no room air circulates into the air space. In other words, if the homeowner attempts to insu-late a window by pressing an insulating sheet against the glass, he "can't miss"; even if, for some reason, the plate is 5/16 in. distant from the glass, the effectiveness of the plate is high.

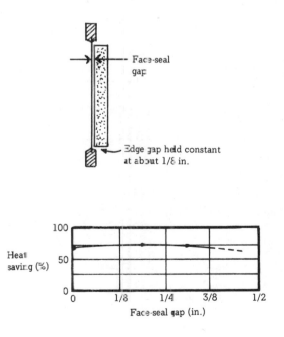

205

## OVERLAP SEAL AGAINST SASH FRAME

In exploring the tolerance on gaps at overlap seals against the sash frame, I used a 1½-in. Styrofoam plate and, in the first test, pressed it firmly against the wooden members of the sash frame. There was then a 1-in. air space between the insulating plate and the glass. In a subsequent test I mounted the plate 1/4 in. from the wooden members. In both tests the vertical edges of the plate were wedged tightly between the jambs and there was a 1/4-in. edge gap at the bottom.

The accompanying graph shows the results. The heat-saving decreased only slightly when there were 1/4-in. gaps at top and bottom. I expect that with 1/8-in. gaps the decrease in heat-saving would have been negligible.

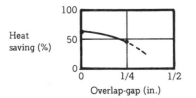

Overlap seal against sash frame

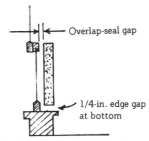

**Effect of cutting central hole in the plate**  Tests made in April 1979 showed that cutting a 3-in. by

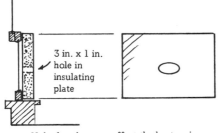

Hole that does not affect the heat-saving

1-in. hole in the center of an insulating plate mounted in the above-specified manner (with or without the 1/4-in. gap mentioned) has no detectable effect on the heat-saving if the window is double glazed and reasonably airtight.

## OVERLAP SEAL AGAINST FIXED FRAME OF WINDOW

I made no reliable tests on the tolerance that applies here. Presumably the tolerance is smaller than in the above-discussed cases, because, with the insulating plate applied to the face of the fixed frame of the window, the thickness of the air space between the plate and the glazing is several inches, i.e., enough to allow greater freedom for circulation of the more-or-less trapped air. Guess: the gap should be kept less than 3/16 in.

## EDGE SEAL

Here also I have no reliable data. Guess: the edge-seal gap should be kept less than 1/8 in.

## INDEPENDENCE OF THE UPPER AND LOWER HALVES OF THE WINDOW

I found that when just the lower half of the window was provided with a 1½-in. Styrofoam plate (or with a 1/2-in. Thermax plate), the upper half being left entirely without insulation, the heat saving at the lower half was the same as if both halves had

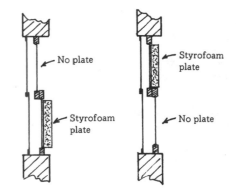

Plate on lower half of window only (at left) and on upper half only (at right).  In such case the plate functions with full effectiveness.

been insulated equally. The converse was true also. In summary, the two halves were found to be independent with respect to use or non-use of an insulating plate. This result was obtained using a fairly airtight window that included main window and storm window with a 3½-in. air space between them.

## UNDERSIZED PLATE PRESSED AGAINST GLASS

How much is the heat-saving reduced if an insulating plate that is to be pressed against the glazing is somewhat too small, i.e., not quite wide enough and not quite high enough?

To answer this question, I made a test in which the 1½-in. Styrofoam plate was considerably undersized: it lacked 2 in. in width and 2 in. in height. When it was pressed against the glass, there was a 1-inch-wide area of exposed glass at the top and bottom and also at the left and right. In a second test, this plate was mounted 3/8 in. from the glass.

As shown by the curve in the accompanying graph, the heat-saving was almost identical in the two tests and was only about 15 or 20% less than would be expected if a full-area plate had been used. The conclusion is that no noteworthy harm is done if a plate (pressed more or less closely against the glass) is a few percent undersized. The harm is merely proportional to the shortfall in plate area.

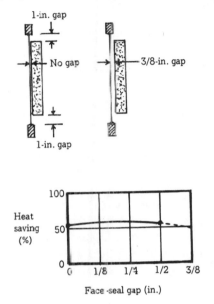

# FLOW OF ENERGY FROM A HOTTER FLAT SURFACE TO A NEARBY COOLER FLAT SURFACE

Designers of thermal shades make much use of thin regions of trapped air. Often they employ thin sheets (of aluminum, e.g.) that, with respect to far-IR radiation, have high reflectance and low emittance.

In this appendix I derive the basic equation for flow of radiant energy from one large flat sheet to a nearby parallel sheet, with a thin region of trapped air between. Also I discuss combined flow: simultaneous flow by radiation and other processes.

The subject is complicated, and not enough reliable information is available. The only bright spot is that the laws governing the flow of radiation are highly accurate and fairly easy to understand.

## FLOW BY RADIATION

### Derivation of Basic Equation

Here I derive the basic equation for flow of radiant energy across the gap between two parallel flat surfaces, for example between two flat shades 1 in. apart, or between a flat shade and a flat sheet of glass 2 in. away. The equation applies to any two flat homogeneous surfaces that are parallel to one another provided (1) the material between them (air, ordinarily) has 100% transmittance for far-IR radiation, and (2) the distance between the sheets is small compared to their width and height.

I call the cooler surface Surface 1 and the hotter one (assumed just 1 F degree hotter) Surface 2. The respective emittances for far-IR are $\epsilon_1$ and $\epsilon_2$ and the respective reflectances for far-IR are $(1 - \epsilon_1)$ and $(1 - \epsilon_2)$.

The basic physical fact is that each square foot of such surface emits, per hour, $\epsilon CT^4$ of radiant energy. C is $1.71 \times 10^{-9}$. $T^4$ is the fourth power of the absolute temperature; for example, if the tem-

perature is 70°F, i.e., 529.6 on the absolute (Rankine) scale, then $T^4$ is $(529.6)^4$, that is, $7.87 \times 10^{10}$.

First I deal with the gross flows X and Y from each surface toward the other. Then I find the difference: the net flow.

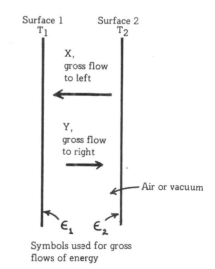

Symbols used for gross flows of energy

Common sense indicates that the gross flow from Surface 2 to Surface 1 is:

$$\begin{pmatrix} \text{Amount emitted by} \\ \text{Surface 2} \end{pmatrix}$$
$$+ \begin{pmatrix} \text{Gross} \\ \text{counterflow} \end{pmatrix} \times \begin{pmatrix} \text{Reflectance of} \\ \text{Surface 2} \end{pmatrix}$$

i.e.:     $X = \epsilon_2 CT_2^4 + Y(1 - \epsilon_2)$

It is obvious also that the gross flow from Surface 1 to Surface 2 is:

$$\begin{pmatrix} \text{Amount emitted by} \\ \text{Surface 1} \end{pmatrix}$$
$$+ \begin{pmatrix} \text{Gross} \\ \text{flow X} \end{pmatrix} \times \begin{pmatrix} \text{Reflectance of} \\ \text{Surface 1} \end{pmatrix}$$

i.e.:     $Y = \epsilon_1 CT_1^4 + X(1 - \epsilon_1)$

To simplify this pair of simultaneous equations, I eliminate $T_2$ by expressing it as $(T_1 + 1)$ and I express $T_2^4$ as $(T_1 + 1)^4$—which, if $T_1$ is extremely large compared to 1, may be reduced to $(T_1^4 + 4T_1^3)$.

The equations become:

$$X = \epsilon_2 CT_1^4 + \epsilon_2 C4T_1^3 + Y(1 - \epsilon_2)$$
$$Y = \epsilon_1 CT_1^4 + X(1 - \epsilon_1)$$

Solving this pair of equations for the gross flows, I obtain:

$$\text{Gross flow to left} = X = CT_1^4 + \frac{\left(\dfrac{1}{\epsilon_1}\right) 4CT_1^3}{\left(\dfrac{1}{\epsilon_1} + \dfrac{1}{\epsilon_2} - 1\right)}$$

$$\text{Gross flow to right} = Y = CT_1^4 + \frac{\left(\dfrac{1}{\epsilon_1} - 1\right) 4CT_1^3}{\left(\dfrac{1}{\epsilon_1} + \dfrac{1}{\epsilon_2} - 1\right)}$$

Subtracting, to obtain the *net flow*, I arrive at this famous equation:

$$\text{Net flow} = (X - Y) = \frac{4CT_1^3}{\left(\dfrac{1}{\epsilon_1} + \dfrac{1}{\epsilon_2} - 1\right)}$$

Btu per square foot per hour,
with temperature difference of 1 F degree

Note concerning the thickness of the gap between the two surfaces: The gap thickness is irrelevant to the flow of radiation. Whether the gap is 1/2 in., or 3 in., the flow is identical, provided that the widths and heights of the surfaces are very much greater than 3 in. The only relevant quantities are the temperatures and emittances of the surfaces. [Strictly speaking, the reflectances too are important; but a reflectance can be expressed as $(1 - \text{emittance})$ and I have written the equation in such a way that emittances, but not reflectances, appear explicitly.] Of course, if the space between the surfaces were filled with black smoke, or black liquid, the situation would be entirely different: the flow would then depend strongly on the gap thickness.

The quantity

$$\frac{1}{\left(\dfrac{1}{\epsilon_1} + \dfrac{1}{\epsilon_2} - 1\right)}$$

is often called the *effective emittance*, E, of the pair of surfaces. The main equation may be rewritten thus:

Net Flow by Radiation $= 4ECT^3$.

or, in general, $\Delta ECT^3 (\Delta T)$.

**Illustrative Examples**

**Example 1**   What is the net flow when $\epsilon_1 = \epsilon_2 = 0$, i.e., when both surfaces are perfect reflectors?

**Answer**   Setting $\epsilon_1$ and $\epsilon_2$ equal to zero in the main equation, one finds that the denominator is infinitely large. Thus the quantity as a whole is zero.

Net flow $= 0$.

**Example 2**   What is the net flow when $\epsilon_1 = \epsilon_2 = 1$, i.e., when both surfaces are perfect emitters and absorbers, and $\Delta T = 1$?

**Answer**   Setting $\epsilon_1$ and $\epsilon_2$ equal to 1, one finds that the denominator has the value 1. Thus the quantity as a whole is simply $4CT_1^3$.

Net flow $= 4CT_1^3$.

If $T_1 = 70°F$, or 529.6 absolute, the net flow is

$$4(1.71 \times 10^{-9})(529.6)^3 = 1.01 \text{ Btu/(ft}^2 \text{ hr °F)}.$$

**Example 3**   Consider the case where $\Delta T = 1$ and $\epsilon_1 = \epsilon_2 = 0.5$. What is the net flow in this case? Here the denominator becomes:

$$\left(\frac{1}{0.5} + \frac{1}{0.5} - 1\right) = (2 + 2 - 1) = 3$$

and the quantity as a whole is:

$$\text{Net flow} = \frac{4CT_1^{\,3}}{3}.$$

If $T_1$ is 70°F, the net flow is 33 Btu/(ft$^2$ hr °F). In other words, cutting the emittances in half reduces the net flow to one third.

## Tabulations

The accompanying tables present illustrative values of the net flow of radiant energy from a flat vertical surface at 70°F to a nearby parallel surface at 69°F. Various values of emittances are used. Radiation resistances values, discussed in a later paragraph, are included also.

| Emittances (pure no.) | | Net flow of Radiant energy (Btu/(ft$^2$ hr °F)) | Radiation resistance ((ft$^2$ hr °F)/Btu) |
|---|---|---|---|
| Emittances are the same, i.e., | | | |
| $\epsilon_1 = \epsilon_2$ | | | |
| 0.0 | | 0.000 | Infinity |
| 0.1 | | 0.053 | 19 |
| 0.2 | | 0.11 | 9.1 |
| 0.3 | | 0.18 | 5.5 |
| 0.4 | | 0.25 | 4.0 |
| 0.5 | | 0.33 | 3.0 |
| 0.6 | | 0.43 | 2.3 |
| 0.7 | | 0.54 | 1.9 |
| 0.8 | | 0.67 | 1.5 |
| 0.9 | | 0.82 | 1.2 |
| 1.0 | | 1.01* | 0.99* |
| $\epsilon_1$ | $\epsilon_2$ | | |
| 0.2 | 0.2 | 0.11 | 9.1 |
| 0.2 | 0.4 | 0.15 | 6.5 |
| 0.2 | 0.6 | 0.18 | 5.7 |
| 0.2 | 0.8 | 0.19 | 5.3 |

| Emittances (pure no.) | | Net flow of Radiant energy (Btu/(ft$^2$ hr °F)) | Radiation resistance ((ft$^2$ hr °F)/Btu) |
|---|---|---|---|
| Emittances are the same, i.e., | | | |
| $\epsilon_1$ | $\epsilon_2$ | | |
| 0.4 | 0.2 | 0.15 | 6.5 |
| 0.4 | 0.4 | 0.25 | 4.0 |
| 0.4 | 0.6 | 0.32 | 3.2 |
| 0.4 | 0.8 | 0.36 | 2.8 |
| 0.6 | 0.2 | 0.18 | 5.7 |
| 0.6 | 0.4 | 0.32 | 3.2 |
| 0.6 | 0.6 | 0.43 | 2.3 |
| 0.6 | 0.8 | 0.52 | 1.9 |
| 0.8 | 0.2 | 0.19 | 5.3 |
| 0.8 | 0.4 | 0.36 | 2.8 |
| 0.8 | 0.6 | 0.52 | 1.9 |
| 0.8 | 0.8 | 0.67 | 1.5 |

* The fact that this number is almost (but not quite) 1.00 is a coincidence stemming from the definitions of Btu, foot, and hour. The numbers are not ratios but absolute physical amounts.

Inspection of the tables reveals these interesting facts:

When both emittances are small, the relative amount of radiant energy flowing is much smaller yet. For example, if each emittance is 0.10, the energy flow is only 0.05 times the maximum flow.

When one emittance is large and one is very small, the latter governs. Thus if the emittances are 0.1 and 0.9, the net flow is only about 0.1 times the maximum flow.

When both emittances are very large, the relative flow is only slightly less than the smaller of the emittances. Thus if both emittances are 0.8, the net flow is 0.67 times the maximum amount.

### Actual Values of Far-IR Emittance

The amount of reliable information readily available on the actual far-IR (4-to-40 microns) emittances of materials used in windows, shutters, shades, and room furnishings is very small. Some information may be found in the ASHRAE 1977 *Handbook of Fundamentals*, p. 22.11 and in *Infrared Systems Engineering* by R. D. Hudson, Wiley Co. (1969).

Some representative values are:

|  | Far-IR emittance |
| --- | --- |
| Aluminum | |
| Foil, dull side | 0.030 |
| Foil, shiny side | 0.036 |
| Sheet, regular | 0.09 to 0.12 |
| Sandblasted | 0.21 |
| Anodized | 0.77 |
| Steel | |
| Stainless (18-8) | 0.44 |
| Stainless (18-8) buffed | 0.16 |
| Galvanized, bright | 0.25 |
| Silver, polished | 0.03 |
| Tin plated onto steel | 0.07 |
| Copper | |
| Polished | 0.05 |
| Heavily oxidized | 0.78 |
| Glass | |
| Ordinary | 0.84 |
| Polished plate | 0.84 |
| Fiberglass batt | 0.75 |
| Common building materials, e.g., wood, paper, concrete, brick, plaster, ordinary paint | 0.9 |
| Miscellaneous: sand, soil, wet soil, water, human skin | 0.90 to 0.95 |
| Foam-type insulating materials | 0.90 (per my guess) |

## Concept of Resistance to Flow of Radiation

Engineers normally draw an analogy between (1) flow of radiant energy from one large flat surface, via a region of air, to a nearby parallel flat surface, and (2) flow of thermal energy through a slab of solid opaque material. Consider first a 2-inch-thick slab of Styrofoam. Suppose that the two surfaces differ in temperature by 1 F degree. Then the amount of heat that flows is about 0.1 Btu/ft$^2$ hr °F). The reciprocal of this, i.e., 10 (ft$^2$ hr °F)/Btu, is called the *conductive resistance*.

Consider now two parallel flat surfaces (at 70°F and 69°F) with an air gap between them.

Suppose that the emittances are 1.0. Then, as explained in previous paragraphs, the flow by radiation is 1.01 Btu/(ft$^2$ hr °F). The reciprocal of this is 0.99 (ft$^2$ hr °F)/Btu. Engineers like to call this the *radiation resistance* $R_r$ of the pair of surfaces and the intervening gap.

In the general case, the reciprocal of the above-derived main equation is called the radiation resistance of the pair of surfaces and intervening gap. That is, whatever the surfaces consist of—whatever the emittances—the term *radiation resistance* is applied to the reciprocal quantity:

$$\text{Radiation resistance} = R_r = \frac{\left(\dfrac{1}{\epsilon_1} + \dfrac{1}{\epsilon_2} - 1\right)}{4CT_1^3}.$$

To a physicist, such terminology may be offensive, because radiation traveling through a vacuum or air encounters virtually no resistance. If, in flow-by-radiation situations, there is anything truly analogous to resistance, it resides in the surfaces themselves, i.e., in the detailed process of emitting and absorbing radiation.

Note that the radiation resistance varies with the temperatures of the surfaces. The hotter they are, the more energy flows and the lower the radiation resistance.

**General equation**  In general, two parallel surfaces may differ in temperature by an amount $\Delta T$ that may be much larger than 1 F degree. The general (approximate) equation that applies is:

Net flow by radiation

$$= 4ECT^3(\Delta T)$$
$$= 3.85 \times 10^{-9}ET^3(\Delta T) \text{ Btu/(ft}^2 \text{ hr)}$$

where $T$ is the *average* Rankine temperature of the two surfaces and $\Delta T$ is the temperature difference.

## FLOW BY RADIATION, CONVECTION, ETC., IN PARALLEL

If there is air in the space between the two parallel surfaces, two kinds of flow occur simultaneously: flow by radiation and flow by ordinary convection. They occur independently. If there were no air in the intervening space (i.e., if there were a vacuum there), the radiant flow would continue as before, but there would be no convective flow. If there were air between the two surfaces but the emittances of the surfaces were somehow made to be zero, as by some ideal silvering, the radiative flow would cease but the convective flow through the intervening air would continue.

The simultaneous flows by radiation and convection are called parallel flows because each starts at the same surface (Surface 2) and ends at the same surface (Surface 1) and the flow mechanisms are independent.

Because the flows are in parallel, the combined conductance is easily found, being simply the sum of the individual conductances. The actual total energy flow with any given temperature difference across the system is the product of the total conductance and the temperature difference.

To find the combined resistance, one merely obtains the reciprocal of the combined conductance.

**Example**  Consider two parallel surfaces, at 70°F and 76°F, with an intervening region of trapped air. Assume that each surface has an emittance of 0.8. Assume that the intervening region of air is thin enough to have a purely thermal conductance of 2. How much energy will flow?

**Answer**  The radiation conductance (with emittances of 0.8) is 0.67. The thermal conductance is

2.0. Thus the total conductance is 2.67. The temperature difference is 6°F. Thus the total rate of energy flow is $6 \times 2.67 = 16$ Btu/(ft² hr).

The accompanying table shows the combined conductance values (and combined resistance values) of a pair of parallel vertical surfaces with an intervening region of air—for various values of effective emittance and various thicknesses of air space. In each case the average temperature of the system is 50°F and the temperature difference across the system is 30 F degrees; thus $T_1$ and $T_2$ are 35°F and 65°F. (Note: When $T_1$ and $T_2$ are much lower, say 0 and 30°F respectively, the combined resistance is about 10 to 25% greater because the thermal conductance of the air is less.)

| Effective emittance E | Thickness of air space | | | |
|---|---|---|---|---|
| | 0.5 in. | 0.75 in. | 1.5 in. | 3.5 in. |
| | Combined conductance, Btu/(ft² hr °F) | | | |
| 0.05 | 0.41 | 0.36 | 0.41 | 0.39 |
| 0.20 | 0.54 | 0.50 | 0.54 | 0.53 |
| 0.50 | 0.81 | 0.77 | 0.81 | 0.80 |
| 0.82 | 1.11 | 1.11 | 1.11 | 1.11 |
| | Combined resistance, (ft² hr °F)/Btu | | | |
| 0.05 | 2.46 | 2.77 | 2.46 | 2.55 |
| 0.20 | 1.84 | 2.01 | 1.84 | 1.89 |
| 0.50 | 1.23 | 1.30 | 1.23 | 1.25 |
| 0.82 | 0.90 | 0.90 | 0.90 | 0.91 |

Source: ASHRAE *Handbook of Fundamentals* 1977, p. 22.12.

Inspection of the data suggests that:

Decreasing the effective emittance from 0.82 to 0.05 increases the combined resistance greatly and reduces the combined conductance greatly.

Changing the airfilm thickness over a wide range (0.5 to 3.5 in.) has practically no effect on the combined resistance or conductance.

The accompanying graphs make the tabulated data easier to grasp. In preparing the graphs I have assumed (guessed) that the thermal conductance (and combined conductance) increases rapidly when the distance between the surfaces is reduced from 1/4 in. to smaller values.

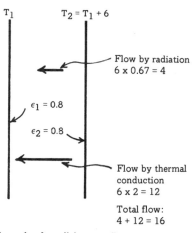

$T_1$    $T_2 = T_1 + 6$

Flow by radiation
$6 \times 0.67 = 4$

$\epsilon_1 = 0.8$

$\epsilon_2 = 0.8$

Flow by thermal conduction
$6 \times 2 = 12$

Total flow:
$4 + 12 = 16$

Example of parallel energy flow between two surfaces

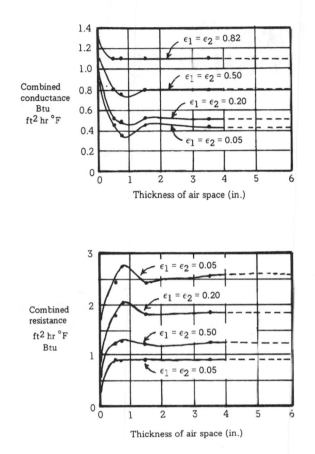

Thickness of air space (in.)

## FLOW WHEN A THICK OPAQUE PLATE IS INVOLVED ALSO

If a pair of parallel surfaces and intervening air gap is in series with an ordinary insulating plate, one finds the total resistance merely by adding the two resistances. The total conductance is the reciprocal of this. The energy flow is the product of the temperature difference and the total conductance.

**Example**   Consider two parallel surfaces (with 1.5 in. of air between) and, immediately adjacent to them, an R-10 Styrofoam plate. Suppose that the individual emittances of the surfaces are 0.67, with the consequence that the effective emittance of the pair is:

$$E = \frac{1}{1/067 + 1/067 - 1} = 0.50.$$

Then one finds from the table that the combined resistance of the pair of surfaces and intervening air is 1.23. Inasmuch as the resistance of the Styrofoam plate is 10, the overall resistance is 11.23. The reciprocal of this, i.e., the overall conductance, is about 0.09. Therefore the energy flow is merely the product of the overall temperature difference and 0.09 Btu·(ft² hr °F).

## FLOW PERTINENT TO A SINGLE SURFACE

This subject is discussed in Chapter 2.

A further fact concerning an outdoor airfilm flanking a vertical wall or vertical sheet of glass is provided by a National Bureau of Standards report "Retrofitting Existing Housing for Energy Conservation: an Economic Analysis," by S. R. Peterson, Dec. 1974. 70 p. SD Cat. No. C13-29/2:64. $1.35. R for surface and airfilm depends on the outdoor windspeed approximately according to this formula:

$$R = \frac{4}{8 + \left(\dfrac{\text{actual windspeed}}{1 \text{ mph windspeed}}\right)}$$

This implies that for windspeeds of 0, 5, 10, 15, 20, and 40 mph the R-values are 0.5, 0.31, 0.22, 0.17, 0.14, and 0.08 respectively.

# TRADE NAMES OF SHUTTERS AND SHADES

Listed here are the names of complete shutters and shades and the companies that manufacture or sell them. The addresses of the companies are given in Appendix 9. By referring to the index one may find the pages on which the products or companies are discussed. The two following appendixes list plates and sheets (rigid or flexible, opaque or translucent) which, when cut to suitable size, may be used as shutters or shades.

| Name of shutter or shade | Name of pertinent company | Name of shutter or shade | Name of pertinent company |
|---|---|---|---|
| Beadwall | Zomeworks Corp. | Silli Shutter | Zomeworks Corp. |
| Blackout | Boyle Interiors Co. | Slimshade | Rolscreen Co. |
| Dimout | Boyle Interiors Co. | Sun Saver | Homesworth Corp. |
| In-sider | Plaskolite, Inc. | Superblind | Boyd, M. D. |
| Insealshaid | Ark-Tic-Seal Systems, Inc. | Therma-Roll | Therma-Roll Corp. |
| Insul Shutter | Insul Shutter, Inc. | Thermafold | Shutters, Inc. |
| Kool-Shade | Solar Science Industries | Thermal Curtain | Thermal Technology Corp. |
| Pella | Rolscreen Co. | | |
| Reversible Piggy Back | Boyle Interiors Co. | Thermo-Shutter | Green Mountain Homes, Inc. |
| Rolladen | American German Industries, Inc. | Warm-In | Conservation Concepts Ltd. |
| | | Weatherguard | Technology Development Corp. |
| Rolling Shutter | Pease Co. | | |
| Rolscreen | Rolscreen Co. | Weatherizer Kit | Plaskolite, Inc. |
| Rolsekur | Rolsekur Corp. | Wind-N-Sun Shield | Boyle Interiors Co. |
| Roman Shade II | Center for Community Technology | Window Quilt | Appropriate Technology Corp. |

# APPENDIX 5

# TRADE NAMES OF INSULATING PLATES AND SHEETS

Here, listed alphabetically by trade name, are the principal insulating materials of interest to persons interested in reducing heat-loss through windows. Included are rigid and flexible devices, opaque and translucent devices, and homogeneous and laminated devices. The previous appendix lists names of actual shutters and shades, and the following appendix lists names of glazing materials. For detailed information on insulating materials, see Chapters 5, 6, and 13

| Name of plate or sheet | Name of pertinent company | Notes |
|---|---|---|
| Air Cap SC-120 | Sealed Air Corp. | 1/8-in.-thick assembly consisting mainly of plastic bubbles 0.4 in. apart on centers |
| Aspenite | Many manufacturers | Like chipboard |
| Astrolon | King-Seeley Thermos Co. | Vacuum-deposited aluminum film between polyethylene films. Formerly called space blanket |
| Beadboard | Many manufacturers | Made of beads of polystyrene foam |
| Cardboard | Many manufacturers | |
| Ceiling panel of fiberglass | Many manufacturers | 1/2-in. panel consisting of fiberglass faced with a a textured, and perhaps perforated, white sheet of paper or vinyl |
| Chipboard | Many manufacturers | Myriad small wooden chips bonded together |
| Cloud Gel | Suntek Research Associates | Sheet that is transparent when cool and opaque when hot |
| Duct board | Many manufacturers | 1- or 2-in. plate consisting of fiberglass faced on one side with aluminum foil and scrim cloth and faced on the other side with white paper |
| Dura-Shade | Duracote Corp. | Thin vinyl sheet with aluminum foil on one side |
| Emergency blanket | King-Seeley Thermos Co. | Low-cost, compactly folding mylar film on which aluminum has been vacuum deposited |
| (Fabrifoil: see Foylon) | | |
| Fesco-Foam | Johns Manville Co. | For use on roofs |

215

| Name of plate or sheet | Name of pertinent company | Notes |
|---|---|---|
| Fiberfill | duPont Co. | Polyester quilting |
| Fiberglas | Owens-Corning | Fiberglass. One type is pink. Available in many forms |
| Fiberglass | Owens-Corning, Johns Manville, Certainteed, Knauf | |
| Fire-X Glassboard | Kemlite Corp. | |
| Foamglas | Pittsburgh-Corning | Glass foam |
| Fome-Cor | Monsanto Co. | Thin layer of polystyrene foam between cardboard sheets |
| Foylon | Duracote Corp. | Also called Fabrifoil Aluminum coated polyester fabric or vinyl sheet |
| Heat-mirror | Suntek Research Associates and Thermofilm Corp. | Transparent sheet that includes a far-IR-reflecting film |
| High-R Sheathing (Kemlite: see Fire-X-Glassboard) | Owens-Corning | Virtually identical to Thermax |
| King-Lux | Kingston Industries Corp. | Pre-anodized aluminum 0.010- to 0.040-in. thick |
| Kynar | Westlake Plastics Co. | Sheet employing vinylidene fluoride |
| Particle board | Many manufacturers | |
| Permafoam | Mid-America Industries, Inc. | Polystyrene foam |
| Plywood | Many manufacturers | |
| Polarguard | Celanese Corp. | Quilt of continuous-filament polyester |
| Sealed Air Solar Pool Blanket (Space Blanket: see Astrolon) | Sealed Air Corp. | 0.3-in.-thick assembly consisting of plastic bubbles 1/2 in. apart on centers |
| Styrofoam (Technifoam: see Thermax) | Dow-Corning Co. | Polystyrene foam. One type is blue |
| Tempchek | Celotex Corp. | Plate of urethane foam faced with asphalt-saturated inorganic sheets |
| Thermalite | Northeast Energy Corp.; also Econ Co.; also Alpha Associates | Honeycomb bubble plastic with adhesive backing |
| Thermax | Celotex Corp. | Plate of isocyanurate foam with admixture of fiberglass. Both faces consist of aluminum foil |
| Thermoply | Simplex Products Group | Plate of very dense cardboard faced on one side with aluminum foil |

| Name of plate or sheet | Name of pertinent company | Notes |
|---|---|---|
| Thermos blanket | King-Seeley Thermos Co. | Vacuum-deposited aluminum film between polyethylene films |
| Thinsulate | 3M Co. | Bat of extrafine fibers of polyolefin. Intended for use in skiers' gloves, jackets, etc. |
| Tyvek | duPont Co. | Sheet of very-tough polyolefin fabric on one side of which aluminum has been vacuum deposited (spun-bonded) |
| Urethane foam | National Gypsum Co. | |
| Zero Perm | Alumiseal Corp. | Aluminum foil between 1/2-mil mylar sheets; used as vapor barrier |
| Zer-O-Cel | National Gypsum Co. | Plate of urethane foam |
| Zonolite | Alumiseal Corp. | Plate consisting of 1/4-in. hardboard faced with Zero Perm |
| Zonolite Styrene Foam | Grace Construction Products, a division of W. R. Grace & Co. | Plate of styrene foam |

# TRADE NAMES OF GLAZING MATERIALS

Here, listed alphabetically by trade name, are the principal glazing materials. Included are rigid and flexible devices, transparent and translucent devices, and homogenous and laminated devices. Materials usually employed as insulators are not included; they are listed in the previous appendix. For detailed information on glazing materials, see Chapter 6.

| Name of plate or sheet | Name of pertinent company | Notes |
|---|---|---|
| Acrylyte SDP | . CY/RO Industries | Double sheet of acrylic plastic |
| Fade-Shield | Madico, Inc. | Fairly transparent (T = 50%) gray-bronze sheet |
| Filon | Vistron Corp. | Translucent sheet made of fiberglass and polyester, sometimes with Tedlar coating |
| In-Sider | Plaskolite, Inc. | Transparent 0.040-in. semi-rigid sheet of vinyl |
| (Kalwall: see Sun-Lite) | | |
| Kynar | Westlake Plastics Co. | Thin sheet of polyvinylidene fluoride |
| Lexan | General Electric Co. | Transparent plate of polycarbonate; available in twin-wall format also |
| LLumar | Martin Processing, Inc. | Thin sheet of polyester film |
| Lucite | duPont Co. | Transparent plate of methyl methacrylate |
| Monsanto 602 | Monsanto Chemical Co. | 0.004-in. and 0.006-in. polyethylene films for greenhouse roof etc. |
| (Plaskolite: see In-Sider, Weatherizer) | | |
| Plexiglas | Rohm and Haas | Transparent sheet of methyl methacrylate |
| Polycarbonate SPD | CY/RO Industries | Double sheet of polycarbonate |
| Poly-Pane | Warp Bros. | Transparent plastic sheet 0.001-in. thick |
| Reflecto-Shade | Madico, Inc. | Slightly transparent (T = 4 to 14%) sheet including extremely thin coating of aluminum |
| Shade-Shield | Madico, Inc. | Moderately transparent sheet (T = 18 to 39%) |

| Name of plate or sheet | Name of pertinent company | Notes |
|---|---|---|
| Solar Membrane | Suntek Research Associates; also Thermofilm Corp. | Transparent sheet that has a far-IR-reflective coating |
| Solatex | ASG Industries, Inc. | Low-absorptance glass |
| Storm Window Kit | W. J. Dennis & Co. | Near-transparent 0.001-in. plastic sheet |
| Sunadex | ASG Industries, Inc. | Low-absorptance glass |
| Sun-A-Therm | ASG Industries, Inc. | Twin-sheet assembly of low-absorptance glass with edges near-hermetically sealed |
| Sun-Lite | Kalwall Corp. | Translucent sheet of polyester and fiberglass |
| Tedlar | duPont Co. | Near-transparent sheet of polyvinyl fluoride |
| Teflon | duPont Co. | Near-transparent sheet of fluorinated ethylene propylene |
| Thermo-Shade | Solar Sunstill, Inc. | Near-transparent sheet, the transmittance of which varies strongly with temperature |
| Twindow | PPG Industries, Inc. | Two sheets of glass with air space between; edges near-hermetically sealed |
| Tuffak | Rohm and Haas | Single sheet of polycarbonate |
| Tuffak-Twindow | Rohm and Haas | Double sheet of polycarbonate |
| Thermopane | Libbey-Owens Ford Co. | Two sheets of glass with air space between; edges near-hermetically sealed |
| Weatherizer | Plaskolite, Inc. | Transparent mylar sheet said to be "semi-rigid" |
| Weatherguard | Technology Development Corp. | Transparent sheet of vinyl plastic |

# MANUFACTURERS OF SPECIAL ADHERING REFLECTING FILMS

One could hardly call a thin adhering film a shade. Yet thin adhering films that have high reflectivity can greatly help reduce energy-loss by radiation.

Some sheets reflect visual-range and near-IR radiation. Some block far-IR radiation also. Some very specal films block just the far IR.

Although such films are, in the main, outside the scope of this book, a little information on such films and the companies producing them is presented below.

*Kalt, Charles W.*, 29 Hawthorne Rd., Williamstown, MA 01267.

Has developed a roll of aluminized mylar that unrolls when electrostatically charged. Several such devices can, when unrolled, cover an entire window. When a window is so covered, practically no visual-range or near- or far-IR can pass through it. When the devices are discharged, they spontaneously roll up, leaving most of the window area open. The change of mode takes only a fraction of a second. Roll diameter may be as small as 1/8 in. (See U.S. Patent 3,989,357 of 1976.)

*Kinetics Coatings Inc.*, PO Box 416, Burlington, MA. Att: Dr. William King.

Has developed an IR-reflecting film which may be applied to the glazing of a window to help keep the building cool in summer. (Ref.: *Windows for Energy Efficient Buildings*, 1, 2 (1979).)

*Sierracin Corp*, 12780 San Fernando Rd., Sylmar, CA 91342.

Has developed electrically conducting plastic film that has a special coating, including oxide and polymer, that increases the transmittance with respect to visual-range radiation. A gold coating provides high reflectance of far-IR. (Ref.: *Windows for Energy Efficient Buildings*, 1, 2 (1979).)

*Suntek Research Associates*, 500 Tamal Vista Blvd., No. 506, Corte Madera, CA 94924.

A manufacturing affiliate is called Southwall Corp.

Has developed a film that strongly reflects far-IR but has relatively high transmittance for visual-range radiation.
(Ref.: *Windows for Energy Efficient Buildings*, 1, 3 (1979).)

(*Thermofilm Corp.*: see Suntek Research Associates)

*3M Co., Energy Control Products*, 3M Center, St. Paul, MN 55101.

Produces many special-purpose optical films. "P-19 Scotchtint Window Insulating Film" has high reflectance with respect to far-IR. (Ref.: *Windows for Energy Efficient Buildings*, 1, 6 (1979).)

*Transparent Glass Coatings Co., Inc.*, 1959 S. La Cienega Blvd., Los Angeles, CA 90034.

Sells "Suntint" film that reflects 59%, absorbs 30%, transmits 11% of solar radiation.

# APPENDIX 8

# AUTOMATIC ACTUATORS FOR WINDOWS, SHUTTERS, AND SHADES

## AMBIENT-TEMPERATURE-POWERED ACTUATORS

Bramen Company, Inc., of PO Box 70, Salem, MA 01970, makes Thermofor, an actuator that provides a stroke of 2½ in. and a force of 100 lb or, with use of a special mechanical linkage, a stroke of 12 in. and a force of 10 lb. A 20 F degree temperature change causes the device to operate. Cost: about $50.

Dalen Products, Inc., of 201 Sherlake Drive, Knoxville, TN 37922, makes SolarVent, which can automatically open a window at 75°F and close it at 68°F. The device has a stroke of 8 in. and produces a force of 9 lb. Cost: about $25.

## SOLAR-POWERED ACTUATOR

Zomeworks Corp. produces a pair of canisters, partly filled with a freon material, that may be attached to a pivoted and well-balanced louver in such a way that the louver will open or close depending on whether the sun is or is not shining brightly. Such devices are used, for example, to control Skylids.

## ELECTRICALLY-POWERED ACTUATOR

Joel Berman Associates, Inc., of 102 Prince St., New York, NY 10012, produces an Electro Shade System that includes a plastic and fabric shade and a reversible electric motor, the latter being within the main roller. A brake is included and may be used to stop the shade at any desired position. Control may be local or remote, manual or automatic. (Ref.: Windows for Energy Efficient Buildings, 1, 11 (1979).)

# ORGANIZATIONS AND INDIVIDUALS INVOLVED

Here are listed some of the main organizations and individuals involved in the development or promotion of devices for reducing heat-loss through large windows. The list includes commercial companies, non-profit organizations, and governmental agencies.

Following the main list is a small list of persons who are working independently.

A recapitulating master index of individuals is presented in the following appendix.

## ORGANIZATIONS

Aardvark and Sons, 167 Webbers Path, West Yarmouth, MA 02673.
Murphy, John S.
Makes sliding shutters.

Alcoa Building Products, Inc., Grant Bldg., Pittsburgh, PA 15219.
Makes windows, including double-glazed replacement windows.

Alpha Associates, 2 Amboy Ave., Woodbridge, NJ 07096
Makes honeycomb-pattern insulating plastic sheets called Thermalite.

Alternative Sources of Energy, Rt. 2, Box 90A, Milaca, MN 56353.
Publishes periodical that has contained articles on thermal shades.

American German Industries, Inc., 14611 N. Scottsdale Rd., Scottsdale, AZ 85260.
Sells Rolladen shutters.

Andersen Corp., Bayport, NY 55003.
Sells windows, including double-glazed replacement windows.

Appropriate Technology Corp., PO Box 975, 22 High St., Brattleboro, VT 05301.
May, David A.
Lowell, Thomas, Director of Research
Makes roll-up thermal shade called Window Quilt.

Approtech Associates, 12 Mt. Vernon Circle, Asheville, NC 28804.
Langdon, William K., 50 College St., Asheville, NC 28801.
Passive solar heating, Thermal shades.

Arizona State University, College of Architecture, Tempe, AZ 85281.
Cook, Prof. Jeffrey
Solar heating. Thermal shades.

Ark-Tic-Seal Systems Inc., PO Box 428, Butler, WI 53007.
Restle, Joseph W.
Makes roll-up thermal shade.

ASG Industries, Inc., PO Box 929, Kingsport, TN 37662.
Sells Sunadex and Solatex glass.

Atlas Industries, Inc., 9 Willow St., Ayer, MA 01432.
Sells Thermax insulating plates.

Bailey & Weston, 30 Garrison St., PO Box 446, Boston, MA 02117.
Sells wide variety of standard shades at wholesale.

Balem Company, Inc., PO Box 70, Salem, MA 01970.
Makes Thermofor actuator for automatically opening and closing vents etc.

Bear Creek Thunder, PO Box 948, Ashland, OR 97520.
Barnes, Jeffrey L.
Reichmuth, Howard S.
Developed sliding and folding shutters for sloping windows.

Brattleboro Design Group, Box 235, Brattleboro, VT 05301.
Ross, Alan
Designed slat-type roll-up shutter.

(Celotex Corp.: see Jim Walter Corp.)

Center for Community Technology, 1121 University Ave., Madison, WI 53715.
Korda, Randy, Executive Director
Korda, Nancy, (insulation of windows)
Kummer, Susan, (insulation of windows)
Provides plans for multi-layer home-made shutters and shades.

Concerned Low-Income People, Inc., 2001 Valker Rd., PO Box 1836, Minot, ND 58701.
Riemers, Roland
Thermal shutters. Window-type collectors.

Conservation Concepts Ltd., Box 376, Stratton, VT 05155.
Moorhead, Clare F., President.
Sells thermal curtain called WARM-IN Sealed Drapery Liner

Cooperative Extension Association of Clinton County (of New York State), County Court House, Plattsburgh, NY 12901.
Hill, Gerald H., Thermal shutters.
Kipp, Rodney, Thermal Shutters

Cornell University, New York State College of Human Ecology, Dept. of Design and Environmental Analysis, Ithaca, NY 14853.
Cukierski, Gwen
Hull, M. V. R.

Cornerstones, 54 Cumberland St., Brunswick, ME 04011.
Wing, Charles G. Thermal shutters. Low-cost buildings.
Gorham, Jonathan W. Thermal shutters

Dalen Products, Inc., 201 Sherlake Dr., Knoxville, TN 37922.
Sells actuators for opening greenhouse vents etc.

Dayton Corp., 11 Beacon St., Boston, MA 02108.
Foster, Kenneth J.
Ball, Robert J.
Makes acrylic glazing assemblies.

Dennis (W. J.) & Co., Elgin, IL 60120.
Sells Storm Window Kit which includes thin near-transparent sheet.

Domestic Technology Institute, Evergreen, CO.
Lillywhite, Malcolm
Sells solar house and solar greenhouse equipment.

Dow Chemical Co., 2020 Dow Center, Midland, MI 48640.
Produces Styrofoam.

duPont Co., Plastic Products and Resins Dept., Wilmington, DE 19898.
Produces many plastic sheets, including Lucite, Teflon, Tedlar, Tyvek.

Duracote Corp., 350 N. Diamond St., Ravenna, OH 44266.
Sells Foylon, an aluminum coated sheet. Also Dura-Shade.

Econ Co., 286 Congress St., Boston, MA 02210; also 139 Gainesville Rd., Dedham, MA 02026.
Foster, Kenneth J., President
Truog, Richard B., Technical Sales
Sells Thermolite, a honeycomb bubble plastic with adhesive backing. Associated with Northeast Energy Corp.

Egge Research, Box 394B, RFD 1, Kingston, NY 12401.
Bauch, Tamil
Has developed windows somewhat similar to Beadwall.

(Energy Research Center: see Syracuse Research Corp.)

Environmental Research Institute of Michigan, Box 8618, Ann Arbor, MI 48107.
Maes, Reed E.
Greenhouses and thermal shutters.

General Electric Co., Schenectady, NY 12301.
Silverstein, Seth D. (Corporate Research and Development Division.)

Grace Construction Products, a division of W. R. Grace & Co., 62 Whittmore Ave., Cambridge, MA 02140.
Makes Zonolite Styrene Foam.

Green Mountain Homes, Inc., Royalton, VT 05068.
Kachadorian, J.
Sells thermal shutters.

Harvard University, Harvard Business School, Cambridge, MA 02138.
Jackson, Tod
Shurcliff, William A., 19 Appleton St., Cambridge, MA 02138. Honorary Research Associate in Physics Dept.
Solar heating. Thermal shutters and shades.

Helio Construction, 190 E. 7 St., Arcata, CA 95521.
Pryor, Roger
Sells plans for roll-up shade. Formerly called Rainbow Energy Works.

Homesworth Corp., 54 Cumberland St., Box 565, Brunswick, ME 04011.
LeMay, Robert C., Jr., President
Sells "Sun Saver" do-it-yourself thermal shutter.

Illinois Institute of Technology, Dept. of Mechanics, Mechanical and Aerospace Engineering, Chicago, IL.
Issued report "Window Shades and Energy Conservation" in 1974.

Insul Shutter, Inc., Box 338, Silt, CO 81652.
Eriksen, George
Sells thermal shutters for windows and doors.

Insulating Shade Co., Inc., PO Box 282, Branford, CT 06405.
Carlson, Gustaf
Hopper, Thomas P.
Sells 3, 4, or 5-layer roll-up shade assembly.

Jaksha Solar Systems, 420 S. 11th St., Lincoln, NE 68508.

Jaksha, Jerry
Solar heating. Thermal shutters.

Jim Walter Corp., PO Box 22601, 1500 North Dale Mabry, Tampa, FL 33622.
(Celotex Corp. is a subsidiary or affiliate.)
Commercial Development Group is at P.O. Box 20105, St. Petersburg, FL 33702.
Ross, J. E., Manager of Commercial Development of New Products and Materials, was, in 1979, working on a Thermax shutter.

Kalwall Corp., 88 Pine St., Manchester, NH 03105.
Keller, Scott F.
Sells Sun-Lite glazing of polyester and fiberglass.

Kemlite Corp., PO Box 429, Joliet, IL 60434.
Makes Fire-X-Glasboard.

King-Seeley Thermos Co., Thermos Ave., Norwich, CT 06360.
Kehoe, Burt
Makes Astrolon, Astrolar, Thermos Blanket, etc.

Metallized Products Division, 37 East St., Winchester, MA 01890.
Caterino, Ronald
Steves, Robert
Production facility for vacuum deposition of aluminum coatings

Kingston and Sons, PO Box 762, Stockbridge, MA 01262
Kingston, John
Developed indoor sliding shutter.

Kingston Industries Corp., 205A, Lexington Ave., New York, NY 10016.
Makes King-Lux pre-anodized aluminum sheet 0.010- to 0.040-in. thick.

Knauf Fiber Glass, Shelbyville, IN 46176.
Makes fiberglass products, including fiberglass ductboards.

Madico, Inc., 64 Industrial Parkway, Woburn, MA 01801.
Sechrist, Robert
Makes many low-transmittance and high-reflectance shade materials, including *Fade-Shield*, *Reflecto-Shade*, *Shade-Shield*.

Martin Processing, Inc., Film Division, PO Box 5068, Martinsville, VA 24112.

Makes Llumar polyester film that contains UV stabilizer.

Massachusetts Institute of Technology, Cambridge, MA 02139
Tribus, Dr. Myron
Thermal insulation for windows.
Appropriate Technology Group, Rm. E-40-156.
Stiles, James, 123 Oxford St., Cambridge, MA 02138.
School of Architecture
Johnson, Prof. Timothy E.

Miami University, Oxford, OH 45056.
Moore, Prof. Fuller
Has developed many thermal shutters and shades.

Mid-America Industries, Inc., Box 236, Mead, NE 68041.
Makes Permafoam, an expanded polystyrene.

National Center for Appropriate Technology, PO Box 3838, 3040 Continental Dr., Butte, MT 59701.
Hamilton, Blair (shutters, windows)
Shapiro, Andrew M. (shades, greenhouses)

National Gypsum Co., Gold Bond Products Division, Millington, NJ 07946.
Makes urethane foam plates.

Newton-Waltham Glass Co., 104 Pine St., Waltham, MA 02154.
Sells kit for installing additional glass sheet on indoor side of window.

North Carolina State University, School of Textiles, Dept. of Textile Materials and Management, Box 5006, Raleigh, NC 27650.
Buchanan, Prof. David R.
Performance of thermal shades.

North Shore Ecology Center, Highland Park, IL.
Levine, Sol C., President
Sells insulation for windows.

Oak Ridge Associated Universities, PO Box 117, Oak Ridge, TN 37830.
Barnes, Paul R. (thermal shades)
Shapira, Hanna B. (thermal shades)

Owens-Corning Fiberglas Corp., Fiberglas Tower, Toledo, OH 43659.
Technical Center, Granville, OH 43023.
Carlson, Tage C. G.

Design of thermal shutters employing fiberglass. The company sells many fiberglass products. One of these, called High-R Sheathing, is virtually identical to Thermax.

Pease Co. of Indiana, Ever-Strait Division, 2001 Troy Ave. New Castle, IN 47362.
Antaya, Daniel J., Marketing manager.
Makes Pease Rolling Shutter.

Pennsylvania State University, University Park, PA 16802.
Summers, Dr. Luis Dept. of Architectural Engineering. Wrote article in Jan. 1978 Popular Science Monthly on window insulation.
White, Prof. John W. Dept of Horticulture. Wrote article in March, 1978, Solar Age on use of aluminized shades in greenhouses.

Pittsburgh-Corning Corp., 1 Gateway Center, Pittsburgh, PA 05222.
Makes Foamglas.

Plaskolite, Inc., 1770 Joyce Ave., PO Box 1497, Columbus, OH 43216.
Sells In-Sider and Weatherizer kits for insulating windows.

Primax Plastics Co., 1 Raritan Rd., Oakland, NJ 07436.
Sells double-faced corrugated polypropylene plates.

Princeton Energy Group, 245 Nassau St., Princeton, NJ 08540.
Brock, Peter S.

Rohm and Haas Co., Independence Mall West, Philadelphia, PA 19104.
Sells Plexiglas.

Rolscreen Co., Pella, IA 50219.
Stuart, Autyn
Makes Pella Rolscreen shutter.

Rolsekur Corp. Fowler's Mill Rd., Tamworth, NH 03886. Formerly at Duke Bldg., Thornwood, NY 10594.
Small, William, President
Makes Rolsekur shutter.

Saskatchewan Housing Corp., Regina, Saskatchewan, Canada.
Eyre, D.
Scherle, Ken

Sealed Air Corp., 2015 Saybrook Ave., Commerce, GA 90040.
Sells Sealed Air Solar Pool Blanket.

Season-All Industries, Inc., Indiana, PA 15701.
Schmidt, Fred M. (wrote books on windows)
Sells double-glazed windows and storm windows.

Serrande of Italy, PO Box 1034, West Sacramento, CA 95691.
Jones, Richard G., Vice-President
Makes outdoor roll-up shutter employing hollow slats.

Shutters, Inc., 110 E. 5 St., Hastings, MN 55033.
Swanstrom, P. W.
Has Patent 4,044,812 on hinged folding shutter.

Simplex Products Group, PO Box 10, Adrain, MI 49221. Affiliated with Simplex Industries, Inc., 240 E. US-223, Palmyra, MI 49268.
Makes Thermoply.

Solar Central, 7213 Ridge Rd., Mechanicsburg, OH 43044.
Greider, Ronald P.
Developed bead-type insulation.

Solar Energy Components, Inc., 212 Welsh Pool Rd., Lionville, PA 19353.
Jones, J. Paul, President
Makes Thermo-Shade roll-up shade. See also Solar Energy Construction Co.

Solar Energy Construction Co., Box 718, Valley Forge, PA 19481.
Jones, J. Paul, president
Makes Thermo-Shade roll-up shade. See also Solar Energy Components, Inc.

Solar Science Industries, 10762 Tucker St., Beltsville, MD 20705.
Ganther, Tim
Sells Kool-Shade shades employing external louvers.

Solar Sunstill, Inc., Setauket, NY 11733.
Kaiser, Ralph J.
Developed coatings, called Thermoshade, the transmittance of which varies with temperature.

Solar Survival, PO Box 119, Harrisville, NH 03450.
Poisson, Leandre
Developed thermal shutters.

Solpub Co., Box 2351, Gaithersburg, MD 20760. Also PO Box 9209, College Station, TX 77840.
Field, Dr. Richard L.
Sells thermal shutters and shades. One shutter consists of detached plates that are white on one side and black on the other; fire resisting paint is used.

(Southwall Corp.: see Thermofilm Corp.)

Stanford University, Physics Dept., Palo Alto, CA 94305.
Claridge, David E.

State University of New York, Atmospheric Sciences Research Center, Earth Sciences Bldg., Rm. 324, 1400 Washington Ave., Albany, NY 12222.
Mohnen, Volker A.

Sterling Industries, Inc., 26 Emerson Rd., Waltham, MA 02154.
Wills, John
Sells variety of foam-type insulating boards including beadboard, Styrofoam, and urethane foam.

Sun Quilt Corp., PO Box 374, Newport, NH 03773.
Price, Timothy K., President
Makes Sun Quilt for insulating windows.

Sunpower Inc., 48 W. Union St., Athens, OH 45701.
Beale, W. T., President

(Suntek: see Thermofilm Corp.)

Syracuse Research Corp., Merrill Lane, Syracuse, NY 13210.
Energy Research Center
Kinney, Laurence F., Research fellow.
Thermal shutters and shades.

Technical University of Denmark, Lyngby, Copenhagen, Denmark.
Thermal Insulation Laboratory
Esbensen, Torben V.
Institute of Building Design, Bldg. 118
Koch, Søren

(Technology Development Corp.: see Dayton Corp.)

Temp-Rite, Inc., 3934 NE Union Ave., Portland, OR 97212.
Sells transparent vinyl plastic sheets and special plastic locking strips.

Therma-Roll Corp., 512 Orchard St., Golden, CO 80401.
Banbury, John Q., II, President
Makes outdoor roll-up type shutter employing hollow slats.

Thermal Technology Corp., PO Box 130, Snowmass, CO 81654.
Shore, Ronald
Developed self-inflating thermal curtain.

(Thermatech Corp.: see Thermal Technology Corp.)

Thermo-Electron Corp., 110 First Ave., Waltham, MA 02154.
Persons, Robert
Insulation; solar heating.

Thermoblind Insulated Window Shutters, Ltd., 28 Queen St., Huddersfield, Yorkshire, England.
Sunderland, M. V., Marketing Director
Sells indoor folding hinged shutters.

Thermofilm Corp., 500 Tamal Vista Blvd., #506, Corte Madera, Marin, CA 94927.
Brookes, John, Consultant
Tilford, Charles
Preparing to manufacture plastic sheet that has a far-IR-reflecting coating. Associated with Suntek Research Assoc. New name: Southwall Corp.

3M Co., 3M Center, St. Paul, MN 55101.
Makes Thinsulate and a variety of flexible films (absorbing, reflecting, transmitting).

Total Environmental Action, Inc., Box 47, Church Hill, Harrisville, NH 03450.
Anderson, Bruce
Michal, Charles J.
Prowler, Don
Scully, Dan
Wrote books and articles involving thermal shutters.

National Bureau of Standards, Institute for Applied Technology, Center for Building Technology, Architectural Research Section, Gaithersburg, MD 20234.
Crenshaw, Richard W. (window insulation)
Hastings, S. Robert (window insulation)

University of Arizona, Environmental Research Laboratory, Tucson International Airport, Tucson, AZ 85706.
Peck, John F. (window insulation)

University of California, Lawrence Berkeley Laboratory, Berkeley, CA 94720.
Berman, Prof. Samuel M. (windows and shutters)
Selkowitz, Stephen E. (windows and lighting)

University of Kansas, College of Architecture and Design, Manhattan, KS 66506.
Coates, Gary, Asst. Prof.
Solar heating. Window insulation.

University of Kentucky, College of Architecture, Pence Hall, Lexington, KY 40506.
Levine, Richard S., Prof. of Architecture
Has designed self-insulating window called "Sundow."

University of Massachusetts, Mechanical Engineering Dept., Amherst, MA 01002.
Goss, Prof. William
Has designed equipment for measuring heat-loss through windows, measuring airflow, and monitoring moisture build-up.

University of Saskatchewan, Saskatoon, Sask., Canada S7N OWO.
Dumont, Robert S.
Energy conservation in houses.

University of Texas at Austin, Dept. of Home Economics, Austin, TX 78712.
Grasso, Maureen M.
Measured heat-loss through shades.

University of Wisconsin, Dept. of Engineering and Applied Science, 432 N. Lake St., Madison, WI 53706.
Schramm, Don, Program Coordinator, Energy Extension Service

Vistron Corp., Filon Division, 12333 S. Van Ness Ave., Hawthorne, CA 90250.
Sells Filon sheets consisting of fiberglass, polyester, and Tedlar coating.

(Walter...: see Jim Walter Corp.)

Warp Bros., 1100 N. Cicero Ave., Chicago, IL 60651.
Sells Poly-Pane transparent plastic sheets.

Wind-N-Sun Shield, Inc., 464 N. Harbor City Blvd., Melbourne, FL 32935.
Sells metalized liners for curtains.

Window Shade Manufacturers Association, c/o George M. Schlosser of McBride, Baker, Wienke & Schlosser, Oak Brook Executive Plaza, 1211 West 22 St., Oak Brook, IL 60521.
Schlosser, George M., Executive Secretary
Supported 1974 study of thermal shades by Illinois Institute of Technology.

Zomeworks Corp., PO Box 712, Albuquerque, NM 87103.
Baer, Steven C.
Sherson, Michael

## PERSONS WORKING INDEPENDENTLY

Algaier, Arthur J. Co-inventor with J. W. Restle, which see.

Anderson, John J. Jr., 170 Nottingham Rd., Ramsey, NJ 07446.
Developed honeycomb-like, collapsible, indoor thermal shade.

Boyd, M. D., 32 Cowper St., Ainslie, ACT, Australia 2602.

Caskey, David L., 1602 Zena Lona N.E., Albuquerque, NM 87112.
Has worked at Sandia Laboratory. Developed simple shutter of Styrofoam.

Day, Ralph K., 307 W. Harrison Ave., Maumee, OH 43537.
Developed scheme for avoiding moisture build-up within double-glazed window.

Gay, Lawrence, General Delivery, Marlboro, VT 05344.
In 1979 he was preparing, with James Styles, a book on insulation.

Gerderman, Dale B., 1403 5th St., Las Vegas, NM 87701.
In 1978 he invented a thermal seal for top of window curtains. See U.S. Patent 4,167,205.

Hickok, Floyd, 137 Lawson Rd., Scituate, MA 02066.
Wrote book on insulation for houses.

Kalt, Charles W., 29 Hawthorne Rd., Williamstown, MA 01267.
Invented an electrostatically controlled window shade.

Krueger, George R.
Co-inventor with J. W. Restle, which see.

Newlin, Paul, 82 Pond View Dr., Amherst, MA. 01002.
In about 1979 he received a grant from the National Center for Appropriate Technology to develop a low-cost solar greenhouse with thermal shutters.

Newman, John M., 802 Chestnut St., St. Louis, MO 63101.
Designed a sliding shutter.

Restle, Joseph W., 4630 N. 109 St., Wauwatosa, WI 53225. (See also Ark-Tic-Seal Systems, Inc.).
Was co-inventor, with A. J. Algaier and G. R. Krueger, of window shade system covered by U.S. Patent 3,990,635.

Ritter, Sandy, 312 Prospect Ave., Minneapolis, MN 55419.
Architect. Has developed thermal shutters and passive solar heating systems.

Saunders, Norman B., 15 Ellis Rd., Weston, MA 02193.
Has developed variety of solar heating systems and thermal shutters and shades.

Shurcliff, Dr. William A., 19 Appleton St., Cambridge, MA 02138. (See also Harvard University.)
Author of books on solar heated buildings.

Stice, James, RR 2, Roseville, IL 61473.
Invented absorber-plus-storage system employing phase-change material; the system can serve also as a thermal shutter.

Ventresca, Joseph A., PO Box 394, College Corner, OH 45003.
Developed thermal shutters.

Wright, David, Box 49, The Sea Ranch, CA 95497.
Developed thermal shutters.

# APPENDIX 10

# MASTER INDEX OF PERSONS

Here, listed alphabetically, are all of the persons mentioned in the previous appendix. For each, indication is given as to where to look, in the previous appendix, for affiliation, address, etc.

# ANNOTATED BIBLIOGRAPHY

Periodical American Society of Heating, Refrig-
erating and Air Conditioning Engi-
eers, Inc. Various periodicals.

A-560 Arens, E. A., and P. B. Williams. "The
Effect of Wind on Energy Consumption
in Buildings." *Energy and Buildings*,
1, 77 (1977).

B-245 Berman, S. M., and S. D. Silverstein.
*Energy Conservation and Window Sys-
tems*, Springfield, VA 22151: NTIS,
1975. 100 p. Stock PB-243117. $5.25.

C-380 Claridge, David. "Window Management and Energy Savings." *Energy
and Buildings*, 1, 57 (1977).

— (Crenshaw, R. W.: see "United
States...")

D-115 Dean, Edward, and A. H. Rosenfeld.
"Modeling Natural Energy Flow in
Houses" *Energy and Buildings*, 1, 19
(1977).

D-170 Diamant, R. M. E. *Insulation Desk-
book*. Surrey, England: Heating &
Ventilating Publications, Ltd., 1977.
(Faversham House, 111 St. James Road,
Croydon, Surrey, CR9 2TH, England.
Paperback. 185 p. About $12. Excellent
book employing SI units.)

E-60 Eccli, Eugene. *Low-Cost, Energy-Effi-
cient Shelter*. Emmaus, PA: Rodale
Press, 1976. 480 p. $10.95

Periodical Elsevier Sequoia S. A., *Energy and
Buildings*. (A quarterly published from
PO Box 851, CH-1001, Lausanne 1,
Switzerland. First issue was in May
1977. Annual subscription: $62.50.)

G-90 Gay, Lawrence, and J. Stiles. *The Complete Book of Insulation*. (To be ready
spring of 1980.)

G-140 Gerhard, Geoffrey. "Passive Solar
Retrofit." *Proceedings of the Second
National Passive Solar Conference*,
March 1978, Vol. 1, p. 228.

G-590 Grasso, M. M., and D. R. Buchanan.
"Roller Shade System Effectiveness in
Space Heating Energy Conservation,"
Paper No. 2520, G1A-39, submitted for
publication in ASHRAE *Transactions*
85, Part 1 (1979). (Shows importance
of using aluminum foil with 1-in. air
space and tight edge seals.)

— (Hastings, S. R.: see "United States...")

H-91h Hickok, Floyd. *Home Improvements for
Conservation and Solar Energy*. P.O.
Box 40082, St. Petersburg, FL 33743:
Hour House, 1977. 152 p. $6.80. Paper.

I-402f International Solar Energy Society,
American Section. *Proceedings of the
Third National Passive Solar Con-
ference*, San Jose, CA 1979. $19.

K-640    Korda, Nancy, and Susan Kummer. "What About Windows? A Report on Thermal Window Coverings With Six Design Options," Madison, WI 53715: Center for Community Technology, 1121 University Ave., 54 p. 1978.

L-80    LaVigne, A. B. "Window Insulation: A Neighborhood Demonstration Project." (See I-402f, p. 900.)

M-94    McGrew, J. L., D. P. McGrew, and G. P. Yeagle. "Integrated Heat-Flow in Windows," Applied Science and Engineering, 1978. 40 p.

N-50    National Association of Home Builders Research Foundation. *Insulation Manual.* 627 Southlawn Lane, PO Box 1627, Rockville, MD 20850. 2nd Ed., 1979. 148 p. $10.

O-505    Olgyay, A., and V. Olgyay. *Solar Control and Shading Devices.* Princeton: Princeton Univ. Press. 1957.

P-130    Peck, J. F. "Insulation, Solar Heating, and Improved Evaporative Cooling," University of Arizona, Environmental Research Laboratory, 1976. 60 p.

R-30    Restle, J. W., A. J. Algaier, and G. R. Krueger. U.S. Patent 3,990,635 of 11/9/76 on two-shade system that provides capability of ducting hot air.

S-40    Schmidt, F. M. *The Window Book.* Season-All Industries, Inc., Indiana, PA 15701. 136 p. $2.95

S-41    Schmidt, F. M. "Windows and Condensation," Season-All Industries, Inc., Indiana, PA 15701. 54 p. $1.95.

S-60    Scully, Dan, D. Prowler, and Bruce Anderson. "The Fuel Savers." Total Environmental Action, Inc. 1975. 60 p. $3.00.

S-65    Searcy, J. Q. "Hazardous Properties and Environmental Effects of Materials (SHAC) Technologies: Interim Handbook." Sandia Laboratories, 1978. Used in Solar Heating and Cooling

(Report DOE/EV-0028; UC-11-59-59c. Available from NTIS under Stock Number 061-000-00207-3 for $8.)

S-85i    Selkowitz, S. E. "Thermal Performance of Insulating Window Systems." Privately distributed. Draft edition. 1979. 45 p.

—    (Selkowitz, S. E.: see also "University of California")

S-145    Shapira, H. B., and P. R. Barnes. "RIB: Reflective Insulating Blinds." (Privately published paper distributed by Oak Ridge Associated Universities in 1979.)

S-235aa    Shurcliff, W. A. *Solar Heated Buildings of North America: 120 Outstanding Examples.* Andover, MA: Brick House Publishing Co., 1978. 310 p. $8.95.

S-235y    Shurcliff, W. A. *Thermal Shutters and Shades: Systematic Survey of Over 100 Schemes for Reducing Heat-Loss Through Large, Vertical, Double-Glazed, South Windows on Winter Nights.* Draft Ed., 1977. 196 p. $12. Distributed by author.

S-235cc    Shurcliff, W. A. *New Inventions in Low-Cost Solar Heating: 100 Daring Schemes, Tried and Untried.* Andover, MA: Brick House publishing Co., 1979. 296 p. $12.

S-255f    Silverstein, S. D. "A Dual-Mode Internal Window Management Device for Energy Conservation. *Energy and Building,* 1, 51 (1977).

U-470z    United States National Bureau of Standards. "Window Design Strategies to Conserve Energy," by S. R. Hastings and R. W. Crenshaw. 1977. 300 p. $3.75. (NBS Science Series 104. Stock No. 003-003-01794-9. SD Catalog No. C 13.29/2:104.)

U-471c    United States National Bureau of Standards. "Simplified Analysis of

Thermal and Lighting Characteristics of Windows: Two Case Studies." NTIS. 1978. 111 p. $6.50.

Periodical University of California. "Windows for Energy Efficient Buildings." Stephen Selkowitz, ed. (Published from Lawrence Berkeley Lab., Univ. of Calif. Program 90-3111, 1 Cyclotron Rd., Berkeley, CA 94720. First issue: Jan. 1979. Free.)

W-40 Wade, Alex, and Neal Ewenstein. *30 Energy Efficient Houses You Can Build.* Emmaus, PA: Rodale Press, 1977. 316 p. $8.95.

W-110 Watson, Donald, and Keith Harrington. "Research on Climate Design for Home Builders." (A 9-p. report of mid-1979. Privately distributed.)

# INDEX